MOVING
THE
NEEDLE

MOVING THE NEEDLE

DAVE MACLEOD

RARE BREED
PRODUCTIONS

Copyright © 2024 by Dave MacLeod

www.davemacleod.com - training advice for climbers

Published by Rare Breed Productions

All rights reserved. No part of this book may be reproduced in any form without written permission by the publisher.

Edition 1.0
Published October 2024

ISBN: 978-0-9564281-5-8

Foreword: Dave Cuthbertson
Book design: Claire MacLeod
Cover design: Claire MacLeod
Original front cover photograph: John Watson
Back cover photgraph: Steven Gordon
Editing and proofs: Deziree Wilson
Photos by Dave MacLeod unless credited

ABOUT DAVE MACLEOD

Dave MacLeod is a professional climber and coach based in the Highlands of Scotland. He has been climbing for 30 years and is recognised as one of the best all-round climbers in the world, having climbed E11 trad, 8C boulders, 9a sport routes, XII mixed and 8b+ free solo. He has made hundreds of first ascents across Scotland and the greater ranges of the world, including the Dolomites, Norway and Patagonia. He is also a respected climbing coach, writing a popular blog since 2006 and hosting a successful YouTube channel. Dave has an undergraduate degree in Physiology and Sports Science, a master's degree in Medicine and Science in Sport and Exercise, and a master's degree in Human Nutrition. He is best known for his hardest trad climbs, Rhapsody on Dumbarton Rock and Echo Wall on Ben Nevis.

CONTENTS

FOREWORD
PAGE 8

INTRODUCTION: THE OTHER SIDE OF THE BULGE
PAGE 18

CHAPTER 01.
TAKING THE PLUNGE
PAGE 28

CHAPTER 02.
ILL-EQUIPPED
PAGE 42

CHAPTER 03.
KELVINISM
PAGE 48

CHAPTER 04.
THE ARROCHAR ALPS
PAGE 54

CHAPTER 05.
NO GOING BACK
PAGE 66

CHAPTER 06.
NO RECESS
PAGE 78

CHAPTER 07.
CRAIG'S WALL
PAGE 92

CHAPTER 08.
HEAVY TRAINING
PAGE 108

CHAPTER 09.
THE IRON ROAD
PAGE 122

CHAPTER 10.
THE FUGUE
PAGE 148

CHAPTER 11.
THE HURTING
PAGE 170

CHAPTER 12.
DUMBY DAVE
PAGE 200

CHAPTER 13.
OFF THE RAILS
PAGE 222

CHAPTER 14.
E11
PAGE 238

EPILOGUE
PAGE 260

FOREWORD
DAVE CUTHBERTSON

Previous: On the second ascent of Dalriada (E6 6b) on The Cobbler, with Ben Lomond behind. © Cubby Images

Left: On the first ascent of The Gathering (E7 6b) on the Cioch, Isle of Skye in 2004. © Cubby Images

I first met Dave MacLeod at the Fort William sports centre, where I had just put up a collection of boulder problems, all listed, graded and marked up on a wall-mounted board for everyone to use. From a distance I watched this young lad, stripped to the waist and wearing a beanie, as was fashionable in the late '90s, rapidly dispense with all but the final problem. I had no idea who he was but enjoyed watching him climb, not with the grace and precision of a '70s master, but with a gritty, dynamic fluidity, always on the move. Intrigued by his technique, I asked for some feedback, to which he responded in a quiet manner and mildly educated Glasgow accent, 'How did you do that last problem?' I replied sheepishly, 'Well, actually, I haven't linked it yet.'

In the evolution of modern climbing, each successive generation gives rise to a visionary, someone capable of taking standards to the next level. An individual not necessarily the strongest nor arguably the best, but in tune with the perceptions of what is possible and prepared to make that commitment. In Scotland's past, a small selection of such luminaries includes the likes of Raeburn, Kellet, Cunningham, Marshall, Smith and Spence. There was something in Dave's make-up, his positive and calm nature, understated confidence and dynamic energy that left me thinking, *Is he the future of modern rock climbing?*

My own debut into the world of Scottish rock climbing had begun modestly in the early '70s, and I like to think that Murray Hamilton and I were at least partly responsible for the dawn of a new era that bore witness to a dramatic rise in grades. From the crumbling delights in the Pentland Hills to literally testing the Water of Leith at the Currie Wa's, not to mention a good chase by the parkies on Salisbury Crags, we followed in the footsteps of Scotland's finest, working our way through architecturally urban 'Auld Reekie' haunts and always looking ahead to bigger and better things. Sure, we broke rules and ruffled a few feathers. But breaking rules is almost as old as climbing itself, allowing our great pastime a chance to breathe and move forward. To describe ourselves as overzealous would be putting it mildly, and within a year I had completed my first new route, somewhat pretentiously named The Beginning. It did indeed mark the beginning of a life as a professional climber, mountain guide and climbing photographer. More first ascents followed, a selection of which are considered to be benchmark testpieces; some have even reached the commendable heights of classic status. However, our contribution could not have been possible without the achievements of preceding generations. There are other influences of course, such as the improved equipment, but more than any other factor, it is the marriage between the legacies of the old guard, the building blocks, and the open minds of a younger rising generation.

Left: Physical Graffiti (Font 6B) on the Home Rule Boulder, Dumbarton Rock. Neil Busby looking on.
© Tim Morozzo

Soon, I had found myself pondering the limits of cutting-edge rock climbing. Unlike my contemporaries, I had not, as yet, embraced sport climbing; at that time considered controversial and by and large seen as a threat to Britain's clean climbing heritage. I tended to agree.

First I needed to find a piece of rock with all the attributes of a modern climb, something futuristic and accessible, not without protection and quick drying. After much deliberation I produced a short list of three possibilities at Glen Coe, Dunkeld and Dumbarton. Against my better judgement, I settled on Requiem, a '60s aid extravaganza and the central compelling line of Dumbarton Rock. And so in the summer of 1983, I laid siege to my ethically-charged experiment: 15 days of small incremental steps - 'three moves a day a route in ten', according to Pete Livesey.

Dumbarton Rock is a true forcing ground in the development of modern British rock climbing, with a unique set of objective dangers and hazards more commonly found on inner city streets. A jumble of massive boulders has long since provided a hang-out for neighbourhood kids, and the cavernous, graffiti-covered boulders provide shelter and acoustics during alcohol and drug-fuelled parties. It is most disturbing when struggling up one's chosen route as a head-splitting anthem reverberates all around. It does not pay to find oneself on the wrong side of these kids, whose stature is more than made up for in numbers. In any case, they always have older and bigger siblings.

When the climb finally succumbed to my dogged determination, I didn't think Requiem was excessively difficult, grading it E7. I certainly had more to give, and the physical and psychological highs and lows I experienced were both thought-provoking and educational. Two further possibilities stood out on the headwall: the lines now taken by Achemine (E9 7a) and Rhapsody (E11 7a). I knew that I was in with a chance technically, but I had neither the guile nor boldness without top rope practice. I had reached a personal threshold, and in many ways I had fallen victim to the restraints of my own ethics. Taking climbing to the next level would require a paradigm shift I wasn't sure I was ready for. Brainwashed by my own romantic notion, I simply couldn't shake my traditional roots and reluctantly, all too aware of the consequences of the rising popularity of sport climbing, I resigned to let the next generation make that decision.

At the turn of the Millennium, injuries instigated early retirement, and letting go - pun not intended - was inevitable. Though not an easy decision, it was cathartic, allowing me to enjoy other aspects of life without the insatiable appetite associated with continuously 'feeding the rat'. Photography could never replace my love of climbing, but it proved to be a great motivator and, from a practical point of view, a source of livelihood. Life behind the lens has created many opportunities, and in this regard it

Left: Falling off Achemine (E9 7a) before the first ascent in September 2001, Claire MacLeod belaying.
© Cubby Images

has been a privilege, both as a friend and photographer, to document the incredible climbs of Dave MacLeod and to witness him grow in strength and character over the last 25 years.

For many climbers, those first two or three years are life-changing, a voyage of discovery and adventure without direction punctuated by naivety, epics and hedonistic fun. As an intelligent, somewhat lost, young boy, uncomfortable within the norms of society, Dave's discovery of a very different and transformative world led to an apprenticeship that was also an exposé, revealing flaws and weaknesses many of us would rather turn our back on.

A move into a Dumbarton flat with his childhood sweetheart and now wife, Claire, marked the beginning of a new chapter and focus in his life. Now something of a rising star and a permanent fixture at Dumbarton Rock, Dave accepted the nickname 'Dumby Dave' with a degree of pride, for the accolade was out of recognition of an unprecedented achievement, that being the first repetition of every boulder problem and sport route in the area. Now 15 minutes' walk away from Scotland's premier urban venue, Claire commented, 'God forbid, I've spent enough time at Dumby holding Dave's ropes, and now I can see the crag from our bedroom window!'

Dave began working the wall above Chemin De Fer (E5 6a). With a projected grade of E9 7a, it was his first major new line and of a level of difficulty hitherto unseen in Scottish climbing. I photographed one of his early attempts, my abseil position providing a most extraordinary angle. Readers should not underestimate the psychological and physical level of commitment required to make such an audacious and terrifying lead. Watching him absolutely committed to the hard bouldery sequence of moves that constitute the crux - a soul-searching distance above his last runner - was both inspirational and harrowing. I found capturing those images a humbling and emotional experience, the culmination of my own contribution and Dave's rise to form. To witness him pushing boundaries built upon my own legacy was the highest accolade a climbing activist could ever wish for. Nearly 20 years on, there have only been five recorded ascents of Achemine, with nobody managing Dave's sequence; nor has there been an onsight.

Over the last 18 years Dave has continued to inspire through his climbing exploits in both summer and winter, not just throughout Scotland but in other parts of the world. In video, lectures and social media, he has opened doors to climbers of all levels and shown that, if you really want to achieve in life, just about anything is possible.

Moving the Needle is his third book; the other two, *Make or Break* and *Nine out of ten climbers make the same mistakes*, embrace his academic background in

Opposite: The Meegies (E5 6a), Creag Dubh.
© Cubby Images

sports science, and while both provide excellent essential reading, they are largely educational. In this regard, *Moving the Needle* could be described as something of a departure, a thoughtful and cleverly crafted autobiography that takes the reader on a frank journey through humble beginnings and Dave's pre-climbing years, and concludes with his historic first ascent of Rhapsody (E11 7a) at Dumbarton Rock in 2006. It is a fascinating insight into the mind of a world-class achiever, and the mechanics behind those achievements.

INTRODUCTION

THE OTHER SIDE OF THE BULGE

Left: Climbing Smith's Route with Andy Turner in 2010 while making a film about the legendary week of first ascents by Robin Smith and Jimmy Marshall on Ben Nevis in 1960. I am leading the original line taken by Smith on the first ascent as the icicle variation above the cave was poorly formed. The tapering ramp above the icicle climbed on my solo ascent is clearly seen above me.
© Hot Aches

With a ceiling of churning cloud lining the iced walls above us on three sides, I couldn't help but imagine that those walls might carry on endlessly into the sky. The north face of Ben Nevis feels bigger on a stormy day, but no matter what we attempted to climb, or how we fared, we would ultimately be going back home tonight to Glasgow and real life. The intensity of a hard ice climb would make that prospect more palatable, and I didn't want to walk off this mountain without carrying that feeling with me. I had come to rely on it. But the prospect of earning it today was now in danger.

Peter wasn't impressed with the conditions and didn't like the look of the ice as we inspected the faces from the great snow chute of Observatory Gully. I'd been climbing a lot more than him over the past year, and our accident at this spot the previous winter was distant in my memory but perhaps less so in his. The cornice at the top of the gully had collapsed on us that day, pulling me off and avalanching the gully. I was carried a thousand feet down, leaving Peter to struggle off the top of the Ben in a state of fright. We were uninjured, but Peter hadn't been back on this north face since, and I didn't blame him for his caution today. Waves of spindrift poured intermittently down the ice faces, inviting the thought that a full-on avalanche could appear behind them, out of the cloud above. I'd seen a huge one while climbing on Comb Gully earlier in the winter. At first, I had thought it was a fighter jet, such was the roar. It was only when the blast wave separated the clouds that I could see the entire slope right of Tower Ridge sweep down and plough its way into Coire na Ciste, filling the lochan at its base. As I watched it crash into the coire floor, I understood how utterly helpless any human would be in its path.

After yesterday's thaw, I was less worried about unstable slopes today, and I tried to persuade Peter that the ice looked okay, but his mind was made up. I couldn't face the trudge back off the mountain without a climb, so I announced I would solo a route

Opposite: Inside the cave below the crux icicle on Smith's Route. © Andy Turner

instead. The ice looked thickest on Smith's Route, and there was a small platform cut in the snow at its foot, a sign of the recent passage of another team of climbers. Climbing solo, I'd be up it quickly and could rejoin Peter at the summit if he carried on up the easy snow of Tower Gully.

Peter asked if I was sure. Only the previous winter, Grade V would have been our limit with a rope between us. I could tell he was concerned that my decision was driven by impatience, but I was at least sure I wanted to make a start. I could climb the first pitch and reverse if things weren't right. This outcome was more likely, although I gave it little thought. I was well used to down-climbing out of trouble because I wasn't good enough to continue upwards. Getting to the top wasn't that important to me. All I wanted was to feel fear, the signal I was approaching my limit.

Peter offered to carry my rucksack and meet me on the summit plateau.

Minus the sack full of rope and ice screws, moving over the steepening snow-ice away from Peter felt light and effortless, more akin to the rock climbing I was used to. I paused at the point where the dull white snow turned to deep blue water ice at the start of the first pitch. *Am I really going to climb Smith's Route without a rope?* I was already ten metres up before I got to grips with the question. By then other questions were creeping in. *Is the ice actually good enough to be here without the rope?* It was excellent in places, soft and chewy from a mild southerly wind the day before. My ice tools sank in with a 'BRRRRR' vibration, and the security prompted me to hop my crampon points quickly up and enter a flow of upward momentum. But then one axe placement broke my rhythm and the ice. A chunk delaminated, exposing bare rock below. I pulled the chunk out and could see a gap between the clean rock and the layer of ice. *Should I ignore this warning?*

I approached the cave, a small alcove before the crux icicle, which hung as a half-drawn curtain across a small rock overhang. In the back, an old belay provided respite from the exposure for roped teams. *Should I wriggle in there now to take a break?* There was no need. Without the faff of having to stop and find protection, I'd raced up the first pitch and hadn't paused for a moment to notice the exposure below. But that exposure would surely come as a shock if I had to clamber awkwardly back out of the little hole in the wall and stare a thousand feet down Observatory Gully. Better to press on, if that was what I was actually going to do. I hooked my tools at the lip of the cave and peered at the hanging curtain of ice barring access to the ramps above, and the questions arrived again.

Who knows what goes through the minds of 'proper' solo climbers? Perhaps they have all their questions answered before they start up their climbs. I doubt it. I reckon

they have their problems too. The icicle looked easier to tackle on the left, but the curtain was thinner here. *Could it break? How warm was yesterday's thaw?* I became aware of Peter's footsteps crunching in hard snow, echoing from Tower Gully out of sight in the cloud. *Am I actually just a bastard, leaving him to carry my rucksack while I indulge this need to be on my limit?* How would he feel right now, probably listening back for the sound of falling ice and falling climber? The sound of his feet in the distance brought regret. Why couldn't I just leave it for today? Every time we climbed together now, I'd persuade him to do one more route, have one more attempt. It would always leave us stumbling out in the dark, late home. Always one more route than made a fun day out. I could tell it grated on him at times; I was coming to realise that going home grated on me.

Dave, forget the questions; that left foot, it's going to slip! I had barely noticed the first couple of moves up the icicle as I listened to Peter crunching through the snow. Now, I had to get a hold of myself. I was stretching away from the foot, crampon pasted on a drip-polished bulb of ice on the lower rim of the cave, delaying having to commit and move it onto the thin, brittle edge of the icicle. *What is the right amount of force to kick the icicle?* Too soft, and the points won't bite; too hard, and the whole thing is gone. I lacked the courage to strike the right balance and opted for too soft. Now, I didn't dare weight the foot properly, and it all started to unravel.

Locked down on my right ice tool with my chin pressed up close to its head, I needed to replace the left, but without the luxury of stability from the left foot. I'd have to hope for a good first-time placement. With a conservative reach, the left tool would go in just below the point where the icicle attached to the ramp from which it hung - the most likely spot for it to break. With a real stretch, I'd maybe get the tool to the ramp itself, but this is where the ice would be at its thinnest. Eyes bulging, I reached for the ramp. The tool bit on something, but it wasn't great. I was too stretched to replace it. *The left foot is going to go; my right arm is pumped. It's falling apart.* Finally I started to engage with the climbing properly. *There's no way I can trust my life to that left tool and remove the right one. Get the left out and replace it. Don't half-ass it.*

Just as I pulled the left tool out, my foot slipped from its half-bite in the icicle. Off balance, I swung out in a 'barn door' motion. Almost facing outwards, I couldn't avoid staring straight down the vast chute of Observatory Gully below. A full shot of fear overrode any remaining self-consciousness. I almost laughed. Fear was what I had come for, but the reality of it is another thing. Later, I would realise that, in this moment, I had leapfrogged past all fear to a new place I hadn't yet been to, where the questioning was bypassed altogether and actions were direct and automatic.

Right: The 'white necktie' of Smith's Route (V,5), Gardyloo Buttress, Ben Nevis. The route is in excellent condition, with the small cave below the icicle completely masked by a curtain of ice.

At the crest of the swing, my left tool whipped back and took a quick bite at the icicle, just long enough to get the foot back in. Immediately I weighted it and locked down further, the head of my right tool lowering from my chin to my collar bone. I swung again, decisively into the thicker ice a few inches higher into the ramp. Swarming onto the ramp, the thoughts came flooding back. *What the fuck, Dave! What the hell are you doing here climbing like that?* Ice climbing should be a solid progression of three points of contact at all times; I should never have moved with only two. I tried to recover some composure with my head pressed into the ice, hiding in between my ice tools. As much as I wanted to berate myself for the serious error on the icicle, I understood that it would be still more serious to do so here, with the rest of the crux pitch ahead.

With the agreement to suspend punishment, I moved on up the tapering ramp, surprised to feel almost immediately light and secure again. Smith's Route was likened in a magazine article I'd read to a 'white necktie', with a triangular funnel of snow hanging above the steep buttress below and the icefall spreading out across ramps from the narrowest point, a bulge which now loomed above me. The ramp leading to the bulge narrowed to the width of my shoulders, forcing me closer to the edge dropping away to my left. Around that edge, spindrift hissed over the bulge in a continuous plume. In one more move, I'd be in its path. Surely I'd be granted a moment of respite to make it through to the easy snow? It didn't work out that way. With both ice tools together at the vanishing point of the ramp, ready to attack the bulge, the stream of spindrift became a torrent. No longer a light and floury plume, it widened and became a river of wet porridge, piling up on my soaking gloves, in the pit of my locked-off elbows, on my hood and down my neck. I shuffled to release the pile now cleaving my chest away from the ice, and felt the bite of my crampons loosen with the disturbance.

Again came the feeling that my control was breaking up. My hands now dangerously numb from the endless spindrift, I felt I had no choice but to reverse a couple of moves to get out of the way and warm them again. While I shook blood into my arms, I watched the neat white curtain of spindrift soar off the edge of the bulge, making contact with the face again a couple of hundred feet below. With space to think, I noticed the wind had picked up and changed direction. It must be collecting the morning's fresh dump of snow across the entire summit plateau of Ben Nevis and steadily pushing it over the edge of the north face cliffs. That torrent would not end any time soon. There was no way I wanted to contemplate reversing the icicle; going up was surely the lesser of two evils. How long could I stay here? An hour? Even here, out of the worst of the spindrift, my hands were getting worse with time, not better.

I began to wish. *All I want is to be on the other side of that bulge.* As soon as I articulated this in my mind, it made no sense. What is on the other side of that bulge? It's all the things I looked forward to getting away from: the trudge off the Ben and back to Glasgow, to connections with people I struggled to make. To a restlessness that would send me right back here to some other piece of ice on this mountain. Modern culture often presented reward and comfort as the purpose of life, sport included. *Is that really my role? Climbing, just to get to the top, with discomfort and risk just the Faustian bargain to be endured along the way?*

For some reason, the short term relief of the crux behind me and the prospect of going home to houses, jobs, Claire or friends didn't even provide a lasting comfort to me, at least not on their own. The crux behind me was the end of the reward I took from a climb, not the beginning. Another ingredient was needed, one that seemed to act as a catalyst for taking comfort both from climbing and from life outside it. That catalyst was a need to lean against some project. The project which had captured my attention for the past three years was to explore how hard I could climb.

Was this a 'man thing'? I had heard people suggest this about climbing, that it was related to a need to achieve or demonstrate ability. Other climbers had referred to solo climbing as 'a young man's game'. But how would I know? I hadn't spent any time around men, though I would soon be 19 and a young man - if I could get around this bulge. Perhaps I was growing into a tendency that was coming to me as an adult. But this made little sense. My dad, when I did see him, didn't seem to share this fixation on projects to such a deep level either. Not now, anyway. Until I had started climbing, at 15, I could let things go, like him. If I couldn't get better at running or playing a guitar, I'd accept it. Either I'd changed when I started climbing, or climbing had drawn a latent aspect of my psyche to the fore. Perhaps it was both.

Maybe it was about courage in the face of risk? I had seen other young men, including climbers, speak of trying to prove their own boldness. I didn't feel this, at least consciously. I didn't expect to be good at sport and knew fine well that I lacked inherent boldness. This had been more than obvious in the school playground. Physical or psychological confidence, the type that would have you look a bully in the eye, could also be used to look a potential ground fall in the eye and then keep climbing upwards. But while this type of self-assuredness might be an input to bold climbing, it did not seem to be the only route to it.

I had also observed climbers who appeared to share the same uncertainties about their ability as me. Some of them appeared depressed. Possibly, their bold climbing had a self-destructive element. It would make more sense if I fitted this profile. Depression was in my family - my father's father had committed suicide. Was my climbing an expression of self-destructive instinct? I don't think so. I had seen the devastation suicide had spread far into the future across generations and, despite frequent negative thoughts, I had vowed early never to go there. Besides, when climbing, I felt free of these thoughts, and they were replaced by a drive to survive and to improve both myself and my life.

Later, I would wonder if both these polar opposite groups of young men found themselves on bold climbs for the same reason: a poverty of sense of purpose that was compatible with their make-up. In modern culture, boldness was becoming unfashionable, often equated with recklessness. Whatever drove me to put myself here on this ice route, leaving behind friends, apparent safety and comfort, ran far deeper than a modifiable cultural construct of what a young man should think or do. I could no more pretend it wasn't there than cancel my need to breathe. I could suspend it briefly, but this would soon feel uncomfortable, then desperate.

Strong as it was, this deep motivation was difficult to define, probably because it arose from an ancient human hardwiring that climbing could only indirectly recreate.

Opposite: With Peter McGowan on Ben Nevis.

Climbing was just the easiest way for me to express it. All I could say with confidence was that I knew if I got over that bulge and out of danger, I'd be under another one as soon as I got the chance.

I moved up again and swung one tool through the curtain of spindrift. The icy particles battered my Gore-Tex jacket, and I paused for a moment, listening for a feeling. Could I sit with the discomfort of this fear for ten moves or so? I didn't really decide to place the next tool, but it went in anyway, both arms now invisible behind the whooshing curtain. Like stepping into cold water, the fear was all in the anticipation, and once committed, it washed away on the white current of snow. Cold blood spread the chill of my numb hands down my forearms. But now my hands had begun to come to life again, fed from arteries thumping with heavy beats. The exit gully appeared over the vertical horizon, and I could see the flow of spindrift for what it was. Concentrated at the apex of the bulge it was fast and heavy, but above, as the gully opened out, it was a shallow, even flow. With each tool placement its force lessened until I could step out onto snow, drop my arms and kick out a little foot ledge to rest.

Breathing hard, blood and warmth returned painfully to my hands, and with it came a serene feeling of bliss and lightness. I eagerly grabbed my ice tools again and raced up the easy gully to meet Peter's smiling face at the cornice, a welcome sight.

'You're officially a nutter,' he announced through a relieved grin.

Being a nutter wasn't what I had set out to achieve at the foot of the climb, but his compliment felt unexpectedly rewarding, and I smiled all the way home to Glasgow.

01.

TAKING THE PLUNGE

Left: The Dumbarton Rock boulders, overlooked by the crack line of Requiem (E8 6b).

An ordinary morning at home, off school sick at age 14, was the last time I can recall the sensation of boredom. Being off 'sick' was an ongoing trend as my secondary school career progressed. The school day brought a sense of dread for me, and felt painful from start to finish. It wasn't the work, which I was indifferent about. It was the kids. I was ill-equipped to deal with the rigours of survival in the social hierarchy there. My lack of physical confidence put me in the worst place possible on that hierarchy, having not yet dropped out of the bottom to become completely detached. I didn't yet know that I was about to have the good fortune to reach that stage, and that boredom would be the catalyst.

The examination regime at school was relentless. Boys were testing each other for physical confidence at any opportunity and could smell fear a mile off. My lack of preparation for these tests was not treated gently and any pupil in this position did well to stay below the radar by any means possible.

The onset of a cough or runny nose would make my heart leap - if I could persuade my mum I was sick. I'd done this enough for the relief of being at home to be offset by the boredom of it. I was too young to relate to the themes played out on daytime TV, and my childhood obsession with learning everything I could about planes and building intricate models of them in card and later in wood had run its course. There was nowhere to go with it without travel and money, and I had access to neither in sufficient measure.

One morning, sitting on the sofa at home, the ache of boredom became particularly acute. Next to the sofa sat a bookcase, filled mostly with my mum's books. I drew one book from the shelf, but not to read it.

It was a road atlas of Scotland, which my nan had been sent as part of her Reader's Digest subscription, though she had no use for it. I recalled a conversation between my mum and my nan, discussing if we knew anyone who might want it. Nan didn't drive and mum drove only to commute to work. My dad had raced motorcycles and loved cars, but mum and dad had separated when I was five and I didn't see him at all for five years. He'd started to visit again after that, but only every other week for a few hours.

Dad had worked as a silversmith at Glasgow School of Art, and one of his side interests was making bodhráns. This is an Irish folk instrument, a large, shallow circular drum made with goatskin. You sit with it on your lap, hand supporting the back of the drum, and play by flicking a double-ended beater which looks kind of like a long bone. Dad even made the beaters with his lathe. I thought the sound and rapid beats of the bodhrán sounded fantastic and yet, as instruments go, it seemed to have one of the lowest barriers of entry for a novice to make a decent sound.

Dad had been involved in the Scottish folk music scene for decades. Pipers, drummers and fiddlers. He was more of an artist than a musician himself, but he loved the folk sessions of Glasgow's pubs and the Highland Games and piping competition circuit. He did once take me along to a piper's club meet in Govan, but never encouraged me to pick up a chanter, bodhrán or even just join him for a pint at the sessions he frequented. I never asked him why. He had brought my much older half-brother, Alan, through the junior piping competition circuit, which is still very strong and well-organised in Scotland.

Alan had been pretty successful as a piper and had played an important part in fusing centuries old traditional piping tunes with high-energy rock and folk music. Pipes, guitars, bodhrán and sometimes voice made a pub-friendly mix, so Alan had spent a lot of time in pubs as a youth. Now, in his 30s, this was starting to have an impact and he looked weathered from many a hangover. Perhaps my dad thought twice before leading me into that world. Perhaps he felt he'd already done his job of bringing through a great musician and had outgrown the touring lifestyle. I'm not sure.

I wanted to play pipes, but it seemed like something far off, too difficult to start. But the bodhrán wasn't. Dad had brought one round as my nan, a good artist, had been helping him with a Celtic design on the goatskin face. The bodhrán was away at nan's, but the beater was sitting on the bookshelf beside me. I took a notion to try it out and figured the back of a book would do for now in place of the bodhrán. I didn't want to mark any of mum's books, but I knew the atlas wasn't on anyone's radar. So I picked it off the shelf and beat out a rhythm for a few minutes on the hardback cover until the motivation faded and the boredom returned. I didn't realise it at the time, but this moment was juncture in my life. I could easily have taken a path of music, but with the road atlas still in my lap, another notion drifted across my mind, one which would lead me in a very different direction.

We'd lived in a traditional Glasgow tenement building, a block away from dad's work at Glasgow School of Art, for the first decade of my life. After separating from my dad, my mum, sister and I had recently moved to a house with a garden on the north edge of the city. As a family, we'd never travelled much around Glasgow except by familiar routes to nan's in East Kilbride. The bus route from there back into Glasgow goes over a big hill looking across the whole city. To me it seemed vast, too vast to know even a portion of it. So I didn't even try. But I did understand that my old tenement near the art school was basically Glasgow's centre. Now, looking at the road atlas, I realised that if someone showed me a map of Glasgow, I'd probably struggle to put a finger on where in the city my new house was.

I opened the atlas. It only took a minute to spot my street and my school and to piece together a few other streets and areas of town I'd explored on my bike. A few days before, I'd been out on my bike and on a whim had cycled off down a road for some time just to see what was there. After a while, I'd realised I wasn't quite sure of the way back. I had enjoyed exploring though, and thought of going further next time. I traced my route on the atlas and immediately noticed that not far beyond my previous cycle, I would have run out of city.

Here, the map turned from orange for urban to green for countryside and was dotted with little triangles with odd-sounding names and a number attached. What were those? I had no idea at first. The map legend told me they were summits and gave their height above sea level. From my memory of looking north across Glasgow while on the bus from nan's, I figured out that I'd seen some of these before: the Kilpatrick Hills, Dumgoyne and the Campsies. I also remembered that through a gap between these two rolling hill ranges, I'd seen a couple of higher conical peaks. By thinking about the angle I'd have been looking from on the bus journey, I worked out that it must have been Ben Lomond. To the north, the map looked very different from Glasgow's tangled mess of streets, with sparse, squiggly roads snaking between countless peaks. Some mountain ranges were bigger than Glasgow itself, without a single road crossing them. I couldn't imagine a space that big without any roads or buildings.

About five miles past my previous cycle, a map symbol indicated a viewpoint and beside the symbol was written 'The Queen's View'. That would be my target for the next cycle. After my 'symptoms' had resolved, I got back on my bike and cycled past the north edge of Glasgow and out into farmland. I rode past a lane signposted for an outdoor centre and recognised its name. Our class at school had gone on a residential

Opposite: Alight here for a life-changing day out. © Claire MacLeod

trip there over several days, though I'd opted not to go. What fun would it have been to get laughed at round the clock with no escape? At the time, the worry that I might be forced to go made my stomach turn. But today, on my own, I was enjoying the wooded landscape and hills and cycled on.

A stand of trees obscured the view ahead as the road approached a bend. As I rounded the bend, the trees opened out to a fantastic vista as the hillside dropped away. I pulled in at what must be the Queen's View. At the foot of the hill was a huge loch, with several wooded islands: Loch Lomond. Beyond it, the entire horizon was filled with distant, rugged hills, with a perfect linear snow line cutting right across them from west to east: the Highlands.

It was a beautiful April day and I had been warm cycling in the sunshine. I hadn't expected to see all these mountains plastered in such a heavy blanket of snow, months after the city winter had given way to spring, and I stood and stared for ages. My dad's tales of travelling around the Highlands, playing pipes and going to Highland Games, came to mind. He painted a picture of long journeys and countless glens, villages and islands he'd visited. Seeing the Highlands in the flesh filled me with such a huge sense of space. The thought of exploring just those mountains I could see from this one spot seemed endless. Yet, sitting prominently in the foreground beside Loch Lomond, I recognised from my memory of the atlas what must be Ben Lomond. The atlas had marked distance points on sections of roads, so I knew that this mountain was only three times the distance I'd already cycled.

That evening, I headed up the street to my local library to see if I could find any books that described these hills. I was surprised to find a whole section on mountains, with one book devoted just to the Southern Highlands, essentially all that I'd seen from the Queen's View. I studied it eagerly and wondered how I might cycle to some of them. Could I get there and back in a day? In most cases, maybe not. Flicking through the pages, I came across an appendix tagged on at the back that was different from the previous chapters, which all described walking routes in prose format. It was a section on 'Climbs at Dumbarton Rock'. A friend of my nan's who was keen on walking had once taken me on the train to Dumbarton Rock as a small kid, and I remembered it as a spiky blob 'that used to be a volcano' sitting strangely in the middle of the River Clyde, with a sprawling ancient castle perched on top of it. There was a long, ordered list of strange names and numbers, which I guessed from the brief descriptions accompanying them were named rock climbs. I didn't know what to make of these. The writing was like a code I didn't understand. But one thing I did know: I could get there easily on the train.

My new house was an old stationmaster's cottage, and some of the trains that went past every few minutes had Dumbarton as their final destination. So the place seemed sort of familiar even though I only had a vague early memory of having been there. I even knew off by heart what time the Dumbarton trains left every hour, since I listened to the station announcements all day long from my house. It wasn't a big leap to decide to go and check it out. The next day, I jumped on the train and 20 minutes later was walking down a long, straight, quiet road from the station towards the hulk of Dumbarton Rock.

The road had an odd feel. On one side, a continuous high orange brick wall ran its entire length. On the other, low-slung warehouses filled with whisky barrels were lined up, with manned security checkpoints at each entrance. Their defences seemed more impervious than the castle on the skyline above. As I walked, the sounds of traffic, trains and deep thuds from a nearby factory faded behind me, giving way to conspicuous quiet. I hadn't brought the guidebook with me in case I lost a library book, but I remembered that the climbs were accessed via a narrow alley at the foot of the rock itself. I reached the entrance to the gloomy alley, which was overgrown with weeds and strewn with rubbish. I entered with my wits about me.

Before we moved to the north of Glasgow, to what I soon learned was one of the better-off suburbs of an otherwise deprived city, I was well used to playing in lanes like this and in areas of wasteland behind city centre buildings. The tenement block of my old house had backed onto Sauchiehall Street, at the time one of the busiest and most 'vibrant' main streets in Glasgow. Sandstone tenement blocks were built all across Glasgow during the Industrial Revolution to support the rapidly expanding population working in heavy industries like shipbuilding. At that time, they didn't have sanitation and so had a shared 'back court' facing away from the street with toilets, a water pump and space for drying laundry. Latterly, when indoor toilets were added, many of these back courts became neglected spaces, except by Glasgow's kids. After the Second World War, Glasgow became very multicultural, although this was, and to an extent still is, concentrated in certain parts of the city. The part where I lived had many Chinese, Pakistani and Indian families. In my first couple of years at school, there was only one other Caucasian and native English-speaking kid in my class, a girl called Jill. The languages of the playground were Mandarin and Punjabi. I spent a lot of time with Jill.

In the back court I played with a small group of boys from mainly Pakistani families. We would carefully climb past some barbed wire-topped walls that had spaces big enough to squeeze through and explore the narrow lanes behind Sauchiehall Street's shops. Mostly we just got a thrill from trying to go as far as we could without being seen, but we had a keen sense of when and where to avoid trouble. Fights on Sauchiehall Street were not uncommon, and it became second nature to steer clear of adults lurking in lanes. The other boys, who were adjusting to a new culture and had very little English, were even more timid than me. One of them only knew how to say 'me bread' - critical language skills when you are going hungry. In the middle of the back court was a huge square shaft that dropped into the ground. A thick slab of iron lay over the entrance, presumably to protect people from falling in. We had no idea where the shaft led and were desperate to find out if there might be treasure down there. The cover didn't sit square over the hole, and with all our strength as seven-year-olds, four of us got our fingers under it and tried to lift it off. It barely shifted an inch.

One day, as we were playing, we stopped and turned as an old VW touring van screeched up the steep hill from Sauchiehall Street, its engine wailing. It jolted to a halt opposite our back court, the side door flew open and five men in balaclavas leapt out and ran towards us. We recoiled in shock, but they sprinted right past. Two of them grabbed the iron slab covering the big hole, flipped it, and one man leapt in. He began tossing out several big bags, which jingled as they plopped onto the grass. We watched, dumbfounded, as the men grabbed a bag in each hand and sprinted back to the van, piling in on top of each other. The van screeched into life before the last man was half in the side door. Desperately grabbing a seat to avoid falling out, he dropped a bag, and as the van sped off, an enormous pile of coins poured all over the street. We ran over and began filling our pockets. We later learned that, the night before, the men had robbed the pub that backed onto our tenement at knifepoint and they'd stashed the Saturday's takings in the hole for a quick getaway. Coming back to retrieve it, they were understandably not keen to linger for long. All the coins were later handed in to the police, except the ones picked up by my uncle, who had walked past on his way to the same pub and picked up enough to cover his first pint or three.

Much as I was familiar with reading places and people in the back streets of Glasgow, my strategy for living in them was mainly to turn around at the first sign of trouble. Despite the barbed wire and secured checkpoints guarding the whisky vaults of Dumbarton, I sensed this place was okay. I couldn't hear anyone in the alley leading around the base of Dumbarton Rock, so I cautiously wandered round. The high brick wall continued on one side, mossy cliffs on the other. But through a couple of collapsed sections, I could see that the walled-off area was a demolished factory of some sort. A swathe of land was buried in a thick layer of twisted, rusting steel rods and lumps of broken masonry.

The alley soon opened out again towards the shore of the Clyde estuary. The outline of a massive boulder came into view, silhouetted against the afternoon sun. It was diamond-shaped and about 40 feet high. A few steps further was an even more impressive sight: a completely smooth, gently overhanging sheet of orange and black streaked rock cut the sky above the boulder. This was 120 feet high and looked like a piece of bold modern architecture from some foreign city far wealthier than Glasgow. Continuing round past the giant boulder, the full height of the wall revealed itself and I got an even bigger surprise. Two rock climbers were dangling right in the centre of the wall, one of them 20 feet above the other. They were positioned on the line of a thin crack that split the centre of the overhanging headwall and progressively faded out in the upper third.

I stopped in my tracks and stared. *This place is crazy.* The climbers wore bright lycra leggings and vests and had lean, wiry but muscled upper bodies. Their rope ran right from the top of the wall and the upper climber seemed to be searching for holds as he hung from a piece of climbing equipment buried in the crack. The combination of a gentle breeze coming off the Clyde and the natural amphitheatre of the cliffs carried their voices towards me, and I stood and listened for a long time, transfixed. The upper climber was jamming his hands in the crack but complaining that he couldn't pull up, relaying to his partner that 'the jams are shite'. He kept asking his partner to haul on the rope, which I figured out was arranged like a pulley above him.

Although they were just hanging on the rope and not sounding too enthusiastic about their progress, I was completely inspired watching them. Just to be in that position on such a smooth, intimidating wall looked fantastic. It was the highest and steepest part of the whole cliff and I could see why they'd chosen it. The challenge of that particular line was just so obvious and natural. The other, easier-looking lines on either side seemed only to underscore the impressiveness of the central crack. I thought to myself, *Whatever that route is, I want to do that.* I cannot really say why my resolution at that moment was so absolute, or why my complete absence of any physical confidence didn't kick into gear and immediately pour cold water on it.

I suppose that is how moments of inspiration are defined, at least in part. The desire, if not need, to do something is burned into your mind regardless of how rational it is. Rationality is simply forced to try and catch up. Ignorance was more my problem at the time. Since I was too scared to bring the library book to the crag, I couldn't remember all the descriptions, so I didn't realise that the crack those climbers were on was a route called Requiem. It had been the hardest route in Scotland for a decade, since it was first free climbed by Dave Cuthbertson in 1983. But to me, it could have been any climb. I had no idea of the gulf of difficulty between this and the adjacent routes.

Right: On Toto (Font 6A+) at Dumbarton Rock.

Left: No.2 Route Direct, (Font 4+) at Dumbarton Rock in 1994. My first piece of climbing equipment was the Troll harness, which was sold in two separate parts by Tiso in Glasgow. I bought the waist loop as soon as I could afford it, but after a few painful abseil practice sessions, saved up for the leg loops.
© Peter McGowan

I did remember one description though: the last route in the book was the rightmost climb on the Rock's basalt cliff. I remembered that it said it started above the water, at a big iron ring that must have been used for docking ships. It was supposedly the easiest climb there, and was called 'Plunge'. As the climbers abseiled down their ropes to the ground to rest, I explored the enormous boulders below the cliff and walked round and saw the iron ring above the Clyde, marking the start of Plunge.

Wanting a closer look, I scrambled across the steep rocks and gullies at the foot of the wall to get to the ring. As I stepped out above the water on smooth, slimy footholds, I could see there was a chance I could slip and fall in, but I was just above the water line and it was only a metre or so deep. At the ring, a diagonal ramp of rock leaned up and right, gradually steepening for 60 feet to where it met the castle wall. The first few moves up the ramp seemed clearly doable, easy-angled on big, flat footholds. So I went upwards on the understanding that I could easily reverse if I needed to. As the rock steepened, it also changed character. I was above the wave-washed rock now, and where it had been polished and slippery from the sea, it now felt rough and grippy, and both feet and hands felt more secure. This led me on higher to the point where I realised I must not fall. But the angle of the wall meant I could stand in balance and take my time. Before I knew what I was doing, I was very high on the wall, feeling exposed and realising that I could no longer rely completely on my feet. But I also felt exhilarated to be there and could see the castle wall was getting closer above. Could I climb a castle?

I reached the castle wall, which was vertical with small edges poking out here and there from the rough rock wall, held together with ancient mortar. I attempted to climb it direct, but after a single move it became apparent this was a very bad idea, and I teetered back down. Feeling rather stuck, I lingered on the sloping ledge below the castle wall for some time, unsure what to do. Eventually it became clear that the only escape was to reverse the way I'd come.

Downclimbing was scary and difficult, but I re-emerged on the grass below Requiem. The two climbers were back on it for another attempt. 'Look, that wee guy didn't get up Plunge,' I overheard one of them say. I didn't think much of this. I was well used to vocal judgement of my sporting performance at school from the other kids. It was, after all, just a statement of fact. I did note the irony in their comment, though. At least I had climbed nearly all the way up my route without a rope. They continued to fail to make a single move on theirs, even with their partner hauling on the rope to assist.

I wondered if Plunge perhaps avoided the castle rampart by moving to its left or

Left: Visiting the sport crags of Glen Ogle for the first time. I didn't yet have a rope, harness or rock shoes, but just tried to climb as high as I could. Here starting up Digital Quartz (8b).
© Barbara MacLeod

right. After milling about awkwardly for a while, I worked up the curiosity to go back to my high point. The traverse out right below the castle had a huge clump of hanging brambles blocking the way, so I guessed it couldn't go that way. The left traverse looked clear, but as I moved around, it dropped away to overhanging rock with few holds. After two more attempts and downclimbs, I gave up.

By now the climbers had left, and without them around to pass judgement, I thought about trying to climb the boulders, which were mostly very steep. I attempted to get off the ground on a couple of routes which had white chalk on them, but could barely even hang on, or simply couldn't figure out how the chalked-up features could even function as holds. My best effort was to get one move up, only to slip off suddenly and land in a heap in the grass and nettles below.

Despite the difficulty of the climbs, I felt very relaxed here and found that, even though the afternoon was getting on, I didn't want to leave. The place felt friendly for two reasons. Firstly, there was no one around to judge my inability to climb these walls. Secondly, once I fell, rolled in the 'jaggies' and picked myself up to stare at the holds again, they were still there, silently waiting for me to try again. There was no pressure, and this was a revelation to me.

Leaving the boulders to head back through the alley to catch the train home, I turned and stared at Requiem for a long time, not wanting to go. This place felt accessible and yet intimidating at the same time. But the intimidation of being 60 feet up Plunge, wobbling around on the castle walls, felt much more manageable than the playground at school. I was utterly ill-equipped to deal with the school playground, and so paradoxically this environment felt more welcoming. To get started, I didn't need money, knowledge, contacts or anything more than basic sense not to go up anything I couldn't climb back down. As I stared at Requiem, it really cemented in my mind that I'd love to do it, no matter how long it would take to develop the skill. It heartened me that the muscled climbers I'd watched trying it for the whole day hadn't been put off by the fact that they could barely pull on a single hold. If it didn't matter to them, perhaps it shouldn't to me either. The rock, after all, would still be there tomorrow.

Right: Climbing at Craigmore with an improvised method for self-belay top roping using a knotted rope. I'd found both the rope and quickdraws, which had been left behind by other climbers at other crags.
© Barbara MacLeod

02.

ILL-EQUIPPED

As soon as I got home, I flipped open the Southern Highlands book again and tried to make sense of the climbs described in it. First, what route had the two climbers been on? Requiem. The hardest route yet free climbed in Scotland and the first route to be given a grade of E7. The route was later considered to be E8 and would have been one of the hardest rock climbs in the world when it was free climbed by Dave Cuthbertson. Oh well, that little dream didn't last long then. I laughed at how ridiculous it was for me to declare that I wanted to climb Requiem, only to find out that it was a completely unobtainable goal. It had been the first time I'd felt a spark of real inspiration about anything. Yet as I let the goal die off in my mind, I noted an uncomfortable dissonance: it would not die completely. Although I pushed the thought of Requiem to the back of my mind, it was already too late to erase it.

I also read the description for Plunge. The route did indeed go rightwards at the castle wall and not to the left, where I'd tried to climb. My failure suddenly didn't seem so bad, and I immediately decided to return the following day to try again. Once I'd squeezed awkwardly past some thorny bushes below the castle wall, the rightwards route led easily up a ramp to a spot where I could reach the top of the castle wall and mantelshelf over onto its ramparts. For a few minutes I sat pleased with myself, enjoying the view and the notion that I entirely owned the success. Aside from the route description, nobody had shown me anything or put a rope above my head. Easy climbing or not, realising that I could figure out how to make progress on my own was a great feeling.

Climbing was quite unlike most other sports I'd seen. Even individual sports like running were carried out in groups, at least at school. You could watch and learn from others, or even use their pace to push yourself. This is good, of course, most of the time. But since I was more accustomed to doing things on my own, I soon noticed

Opposite: Ben Lomond overlooking Loch Lomond.

climbers bouldering alone at Dumbarton Rock and liked the fact that social skills weren't a prerequisite for participating in climbing. Neither was equipment. A pair of rock shoes and a chalk bag make a huge difference to how enjoyable it is to climb. But they aren't necessary to get started.

As I gathered first impressions of the world of climbing, a schism emerged from the pages of another book I'd borrowed from my library, which detailed techniques and equipment for mountaineering. It stressed that travel on Scottish mountains was a serious business and would-be climbers should not only have expertise, but maps, compass, boots and waterproofs. Yet in other books, many of the greatest climbers recounted with reverence experiences when the right equipment or knowledge was not to hand for one reason or another. Ground-breaking new routes, by definition, took new territory where equipment, preparation and execution of the climbs were often improvised. Advances in equipment and expertise *followed* the vision of expert climbers, not the other way round. I struggled to reconcile the apparently opposing approaches of leading climbers, and advice offered to those starting out.

I had been hoping to get a chance to climb a proper mountain as soon as possible, and this warning about doing so without prior knowledge and equipment placed a significant barrier in my way. There was no way I could access any of those things, and so I reeled in my ambitions. My mum had offered to drive me for a day out to the closest peak in the Southern Highlands, Ben Lomond. I could hardly say no, but perhaps I should pour cold water on the prospect of getting to the top? Instead, I could just walk up the lower slopes to see it up close.

Leaving my mum behind at the shore of Loch Lomond on a sunny Saturday morning and climbing up through the wooded lower slopes of Ben Lomond, I looked carefully at others setting off up Scotland's most southerly Munro. All of them were indeed wearing big leather boots; most carried fancy-looking rucksacks, no doubt containing the list of equipment detailed in my mountaineering book. Some even had these more visibly at hand - maps in plastic sleeves, compasses and even whistles dangling from their necks. As the path cleared the tree line and a fine view opened up to Loch Lomond below, I had my first sense of being in mountain territory, albeit ill-equipped. I had the same guilty self-consciousness I usually had when bunking off school: *I shouldn't be here*. I stopped at a boulder and decided to enjoy the view for a while and then go back down. The other walkers continued past. Some of them caught my eye and I sensed their judgement about the trainers and schoolbag I wore.

I carried a thin cagoule I used for school and a glass bottle of Irn Bru, the standard for Glaswegian days out. As I sat and drank my ginger, excitement to be finally here

melted into anti-climax. I'd only been on this mountain for an hour. It wasn't even lunchtime and I didn't want to go down or go home. I wanted to follow the steady stream of people continuing on towards the bulk of Ben Lomond above. I shouldered my schoolbag to walk down just as another walker passed by. He wasn't like the others and had soft white trainers like me and no rucksack. On his back he wore only a large tattoo and a Glaswegian bright pink tan. He carried his equipment in each hand: a cagoule and a glass bottle of Irn Bru. His pace was about twice as fast as everyone else. *He must know what he is doing.* Without hesitation, I jumped up and followed him at a distance, struggling to keep up. He didn't stop once. With jelly legs, burning with fatigue and clumsy on the rocky path, I could only glance upwards occasionally to see if I was falling behind, and so my first sense that we were getting high was a sudden cold wind chilling my sweat-soaked cotton T-shirt. Pausing for the first time, I looked up to see the summit right in front of me.

My mountain leader sat by the cairn and swigged his Bru, and I did the same from a distance. This first summit view was gratifying, with clouds floating by and rolling across the summit ridge, the huge length of Loch Lomond stretching below and a sense of open space I'd never experienced before.

Of all the walkers on Ben Lomond, I had chosen to follow the one who was apparently the least properly equipped. Was this a mistake I should make again? His presence catalysed my first mountain ascent where every other well-equipped walker might easily have catalysed my retreat. Not following their example was also a decision. You certainly have to be careful who you follow, and in my mind, this is exactly what I was doing. Adherence to firm rules about what should or should not be done on mountains is self-evidently wrongheaded. I had years of climbing in front of me before I could confidently see this for myself. For now, a young tattooed lad from Glasgow could show me that it was okay to walk up a mountain on a sunny day without any special equipment.

I was glad, this time at least, to have disregarded the over-generalised advice in my climbing textbook as well as the norms of the other walkers around me. I would need many more nudges in this direction yet before I could disregard much other advice that would get in the way of achieving my naive goal of doing Requiem, or even prevent me from holding onto the goal in the first place. An important first step had been made in the learning and unlearning process.

Of the many things that can be carried up mountains, judgement is possibly the only essential tool. It is useful to bring what you can, but so long as you are open to observing the ground ahead, it will encourage you to exercise judgement and so develop it further. I have often wondered whether it is better for the early acquisition of mountain skills to be led by people or the mountains themselves. There are plenty of exceptions, but it has struck me that many of the best and safest climbers develop their foundational skills without much input from human mentors. They often make dangerous mistakes along the way, but so long as their objectives are not too serious, the consequences may just be a good story to tell, but not to relive.

Amenable adventures undertaken without much prior knowledge or special equipment invite the new climber to begin the long process of learning to observe, anticipate and improvise: three skills that are needed more and more at higher levels of climbing and tend to be those that distinguish average from great climbers. My mentor on Ben Lomond was at the perfect distance from me: visible enough to prevent me from allowing the rigid doctrine of a textbook to get in the way of learning, but distant enough that I had space to start to look at the terrain ahead and think for myself, *What will happen if I go there?*

03.

KELVINISM

Left: On the Glasgow Underground at Kelvinhall.
© Claire MacLeod

In later years it would seem backwards to me that my country would fight passionately to keep hold of its free healthcare service, swamped by the consequences of an inactive population, yet tolerate barriers of expense and exclusivity to one of our most effective forms of healthcare: sports facilities. Thankfully, anyone can still exploit a major loophole; the Scottish mountains remain free if you can get to them. Sometimes it is dark and raining in the mountains though. As a 15-year-old kid visiting Glasgow's best indoor sports facility at the time, I concerned myself with exactly none of these questions. I just wanted to go rock climbing but didn't have the money to get in, at least not regularly. But Glasgow's Kelvin Hall had loopholes, too. Back then, if you paid the few pounds entry fee to climb in the little bouldering wall tucked in the back corner of the building, almost out of sight and out of mind, you received an old-school blue paper ticket. Since very few climbers used the wall compared to the other facilities at the centre, the management hadn't bothered to install any barriers to the wall, or any lockers near it. It was pointed out to me that if any staff ever came to check that you had a ticket, you could simply say you had left it in your locker downstairs and make a run for it from there. More importantly, even the gnarled climbers who had been training here since the '80s had never seen anyone check tickets.

Therefore, as soon as the autumn rain set in during my first year as a climber, I set a pattern of five hours in the Kelvin Hall on Mondays, Wednesdays and Fridays. It was an excellent environment to cultivate something I would need as much as any other attribute as a rock climber: a love for training. Like most kids my age in Glasgow, I was exceptionally weak, unable to manage half a dozen pull-ups at first. Of course, at the time, I actually felt pretty normal in this respect. This is the awful thing about growing up in an unhealthy society. Normal becomes exceptional, and exceptional becomes normal. I was fortunate that the Kelvin Hall replaced the environment which

Right: The Kelvin Hall bouldering wall was made with rocks embedded in the wall. Polished with use, you could climb for hours before wearing out your skin.
© Peter McGowan

I identified with, showing me a new (real) normal where most people were pretty damn strong. Well, maybe the climbers weren't that strong in the 'let me move that fridge for you' sense. But they certainly could pull on some small holds by virtue of apparently not eating very much. Their muscles were fairly weedy, but I could easily see every one through their paper-thin layer of subcutaneous fat.

I started off with several hours of bouldering, using the best of my strength, working the hardest moves I could. As power levels dropped, it was time to head downstairs to the weights gym. Like the climbing wall, this was in a quiet corner at the back of the building and separate from the much more popular 'aerobics' style gym with treadmills and classes. The weights room was probably not the most inclusive place to spend time, but I was comfortable in there. Talking seemed to be avoided unless absolutely necessary. Very large, ripped men agreed swap overs on the squat rack with a code of momentary glances, suggestive pacing and subtle shoulder movements. Very occasionally, someone failed to read the signs and men squared off, while everyone else took a step or three backward. Testosterone hung in the air. There would have been plenty more hanging in their bags in the locker room around

Right: Three winters of climbing on Monday, Wednesday and Friday evenings on the Kelvin Hall bouldering wall built a solid base of finger strength.
© Peter McGowan

the corner. I probably weighed less than the average leg of the others training in there, and so I seemed to fly under the radar. The quiet suited me well and the regulars treated me as the boy I was, just leaving me to my lat pull-downs. I did a lot of lat pull-downs. Mondays, Wednesdays and Fridays were all lat pull-down days. When the big man was finished his squats and I had struggled to tidy up his setup for him so I could access the pull-up bar, I followed up with one-arm lock-offs, one-arm pull-downs and maybe some more lat pull-downs, wide grip this time.

With so many mirrors and so much time resting in silence, I watched the others and the differences between the bodybuilders, powerlifters and athletes became apparent. The bodybuilders looked at themselves in the mirror, always. During the sets, between the sets. The only time I ever saw their eyes drop from the mirror was when the biggest man once let out a high-pitched fart mid-squat - a dangerous moment for anyone who couldn't keep a stony face. Everyone was up to that challenge; they'd had plenty of practice. The powerlifters and athletes used the mirrors only occasionally when doing drills to tighten up form. The rest of the time, their gaze was hard to pinpoint. There was one woman who trained in the gym. She was clearly not a lifter, but I would never have dared ask what her sport was. She stood out among everyone. Her work rate was extremely high - doing some sort of technical drill, power and endurance exercises using body weight or bands on the floor. They seemed never-ending, just the same exercises, session after session.

I came to love the lat pull-down machine. Watching 60kg of weight sliding up and down in front of my face every Monday, Wednesday, Friday. Later, it became 100kg. The place was like a weird silent disco, everyone floating on the endorphins of hard effort, together in the same room but lost in their parallel worlds, never once saying hello. None of us had any idea what journey the other was on while staring into the middle distance between sets. I was happy to be part of this strange band of folk, gradually becoming stronger and paler in the basement of the Kelvin Hall. It's good that people tend to talk to each other more in climbing gyms now. But at that time, I was glad that no one spoke to me and just let me watch my silent mentors and learn the meaning of graft in training.

After three consecutive winters on the lat pull-down machine, I went from five pull-ups to a clean one-arm pull-up. It was one of the first big things I earned fair and square, and I felt proud of it. I'm grateful to my mentors in the Kelvin Hall who helped me learn the enjoyment of training and the most important thing required to make it work: consistency. They taught me this without ever saying a word.

Later in my teenage years, the Kelvin Hall tightened up its entry system. Had I wanted to train in the climbing wall and gym, I'd have needed to pay twice. There was a period where alcohol, being far cheaper, competed for my attention after school, not least because that was what all my friends chose. Thankfully, it was already too late. The loopholes had already engrained a love of training and of being fit and healthy that was burned in, lifelong. If I could wave a magic wand and grant one wish for Scotland, it would be a law that always made a training session cheaper than beer. I cannot think of many investments a country could make in itself that would bring a greater return.

This page: En route to training with Ali Weir and Michael Connor.
© Claire MacLeod

04.

THE ARROCHAR ALPS

Left: The Arrochar Alps have provided steep unclimbed lines for generations of climbers. First ascent of The Cathedral X,11 on The Cobbler, December 2004. © Cubby Images

Like generations of young Glaswegian climbers before me, far-flung climbing destinations like the Alps, the Himalaya or even just the limestone walls of Spain were places we would read about in climbing magazines rather than go to. Those mountains looked impressive in an unreal way. In a sense they might as well have been Photoshop creations for climbing magazines or special effects for a Hollywood movie. We could dream about travelling to them all we liked, but the reality was, we didn't have the money. They didn't really exist in our world, which was limited to where we could afford to walk, cycle or scrape a bus or train fare to.

No one in my family took foreign holidays. My mum had been on a trip to Greece once, well before I was born. But whenever she talked about it, there was always the sense that it was such an unusual and extravagant thing to do. The conversation would inevitably progress to thinking about the more pressing things that money could be spent on. Even though mum had pushed herself hard to provide for us, travel was practically and culturally out of reach, something people did once in a lifetime before returning to the reality of getting by and putting food on the table. I was 16 and had never been abroad, and it never occurred to me that it might be something I could even desire to do. Because of this, I never felt any sense of missing out. It just meant that the mountains I started aspiring to climb were all contained in the real world that applied to me at the time: the Southern Highlands of Scotland.

Thankfully, that world does contain some great mountains. One hour's train ride from Glasgow lies the closest major mountain rock climbing and mountaineering testing ground, the Arrochar Alps. I first read about the climbs in this little range of hills in *Scotland's Mountains*, another book I'd borrowed on repeat from my library and read cover to cover, multiple times. It was essentially about hills and hillwalking, and like almost all the books on climbing in Scotland in my library, it was published by the Scottish Mountaineering Club. As I started to explore Scottish mountain culture through the SMC's books, I noticed that mountaineers seemed quite focused on recording and counting things. The SMC, I later learned, had a reputation for being quite an upper-class (at least relative to us), if not elitist, club. A decent proportion of its members were very well educated, wealthy and either retired or in fields of work where they had spare time to compile information about mountaineering. The club had an archivist and its own library and ran an associated publishing business, producing books of all types about everything to do with the subject. The trust still produces an exhaustive list of climbing guidebooks to this day, for every discipline except bouldering, which they still perhaps haven't caught up with. The books certainly exuded a sense of wonder and deep appreciation for the Scottish mountains. But they did so in the context of a very ordered framework: the layouts of hill ranges and how their topography differed, or how a single climb fitted into the broader history of mountaineering.

On the one hand, I appreciated this sense of order and guidance since it helped me understand what was out there beyond the panorama I'd been struck by while cycling out of Glasgow for the first time. I could put stories of climbing into context and grasp their significance, despite my ignorance as a novice and a lack of any mentors to learn from. It also quenched a curiosity about the Highlands I had from listening to my dad describing them to me as a kid. But never once had he said, 'Right son, jump in the car, we're going to Skye,' so that sense of mystery had just been quietly brewing away, unsatiated. Now, I could see the layout of the hills and appreciate the individual character of each region, not yet having explored them. However, reading the SMC books, I couldn't help but feel that I was peering into a culture that I'd never really be part of. Maybe it was just the language, which was very proper, almost scientific. It reminded me a little too much of the well-spoken teachers at school I felt rather disconnected from.

But the sense of connection strengthened a little as I turned to the final chapter of *Scotland's Mountains*. Tagged on at the back was a history of mountaineering developments, which, on the face of it, fitted nicely into the book. But in the second half of the 20th century, climbing in Scotland had moved well beyond tackling the major ridges and even the rock buttresses of the Highland mountains. The cutting-edge had progressed to higher and higher standards of rock climbing, taking on the challenge of the steepest, most outrageous features. Reading through this history, it struck me that this trajectory had paradoxically brought the most exciting climbs back closer to me, reading the book in Glasgow. Straight away, I turned to a picture of Requiem, the route I'd seen those climbers on at Dumbarton Rock. The image of Dave Cuthbertson lunging between small edges, a couple of moves from the top of the smooth wall, looked all the more incredible since I had seen it in the flesh, with two lycra-clad, muscled climbers barely able to move on it. I could appreciate that what had been done here was a huge athletic achievement. Yet something about the urban landscape of Dumbarton's tower blocks and factories in the background made it seem so much more relatable. The thought of this being played out at Dumbarton, on some random day with people going about their business in the town behind, was so different from how I had previously thought of great sporting feats, normally viewed at a distance, through a TV screen, in a colourful stadium, with Freddie Mercury belting out 'Barcelonaaaaa!'. There was something attractive about the idea that the route was just sitting there, 20 minutes along the train line, and yet was so significant in the world of climbing. I could hang a rope down it myself, touch those same holds and marvel about how impressive a feat it is to climb this wall. The cost to step into the same arena as a world-class athletic performance was a £1.10 half-return train ticket to Dumbarton East - and only if the inspector caught up with me on the train.

Aside from the well-spoken, educated group who were writing these books about Scottish mountaineering history, there was another major cohort, not writing the books, but definitely represented in them through their achievements. This was a working-class army who, for generations, had ventured from the inner cities and headed north to the crags at weekends. An interesting theme had developed over the decades of the 20th century that preceded my entry into the climbing world. Since the 1950s, this working-class half of Scotland's relatively polarised climbing culture seemed better suited to making leaps forward at the cutting edge. This period was the first time climbers started to contemplate moving away from the amenable features of the cliffs - big corners, cracks or other weaknesses - and take on steeper walls and open faces.

At the time, this was a serious business, and the risks of attempting these unclimbed walls were easily as significant as any other form of alpinism, which is saying something. In earlier eras, it was much more likely that if you forged upwards on a new route but ground to a halt because the terrain was too difficult or intimidating, safe retreat was usually possible. If following a crack or corner system, it was easier to get weight on your feet long enough to find a runner to clip to the rope and lower off, or simply downclimb. But as routes progress in difficulty, they become more and more committing, in the sense that the space between upward progress and falling off narrows. If you push on up a steep wall but cannot continue, your forearms are more likely to run out of strength before you can find protection. Moreover, downclimbing becomes extremely difficult. As well as demanding stronger and leaner bodies, harder climbs of this type require either a greater reliance on a judgement call that the section ahead is within your physical limits, or greater boldness to just have a go anyway. Of course, a mixture of all of these attributes is usually required for the hardest climbs, at least in the short term. Younger generations have always pushed the limits of boldness, since it can be called upon in place of fitness or judgement in early years. But sooner or later, usually sooner, a real scare will dial this back to something more realistic, assuming the climber survives the episode. Beyond the naivety of the youngster who has yet to experience how hard the ground is, climbers settle out at different levels of boldness that can then be sustained over years. The various aspects of working-class Glaswegian culture in the mid-20th century seemed conducive to high levels of boldness.

At first, the push towards higher rock climbing standards came from a small club called the Creag Dubh. Among them, Glaswegian John Cunningham and others had reputations as legit hard men who brought their toughness to the cliffs and took on features that previous generations would not. By the time I started climbing, the Creag Dubh seemed better known for its hard drinkers and shared with the SMC a need to attract younger members. A new era seemed to arrive in the late 1970s, with a combination of improved rock climbing equipment and techniques, the adoption of athleticism and specific training and a new generation willing to apply them. What was also interesting was *where* they applied those new skills and techniques.

Many of Scotland's grandest mountain ranges - the Cairngorms, the Torridon hills and even the Cuillin on Skye - tended to form slabby rock faces, less than vertical and well-featured with cracks and holds. The biggest walls could often be climbed at quite amenable grades. This is great if getting into spectacularly exposed places without too much difficulty is the objective. But as *Scotland's Mountains* made clear, through the 1970s and especially the 1980s, rock climbers looking to push the standards deliberately sought objectives that appealed for the opposite reason: they were either grossly overhanging, devoid of cracks and therefore the safety of protection, or they just looked the most intimidating. Moreover, these features seemed overrepresented in the Arrochar Alps, especially on one mountain in particular: The Cobbler.

The West Highland train line from Glasgow to Fort William was made by the Victorians to allow access to the deer hunting forests of the north, among other things. It remains a vital link between the Highlands and the rest of Scotland to this day. As it leaves the Central Belt and the Clyde coastal estuary and snakes up the side of Loch Long, the first visually striking mountain is The Cobbler. It exemplifies the playful and accessible nature of rock climbing. As you stare across the loch at the three spiky peaks that form the mountain skyline, it is obviously small enough to seem inviting to explore, even as a novice. But the overhanging beaks of rock forming the North Peak are remarkable, giving a kind of rugged prehistoric ambience.

Scotland's Mountains described two climbs on The Cobbler's North Peak that jumped out at me. First, there was a picture of Dave Cuthbertson on Rest and Be Thankful, one of the mountain's hardest and boldest routes at the time at E5. Like the photo of Cuthbertson at Dumbarton Rock, it struck me that aside from the crazily exposed position on the rippled dark grey wall, this didn't look like the images of athletes I was used to seeing on TV in mainstream sports. He wore faded shorts and T-shirt, long, grubby white socks, and he had just a few pieces of equipment clipped to his harness. It was clear that, although part of a wider band of young climbers making a jump in standards, Cuthbertson was setting the bar. The book described how another climb on the Cobbler's North Peak, Wild Country, had in 1979 been a huge leap forward in difficulty in the UK. This went directly up the centre of the massive overhanging nook that forms Arrochar's skyline. It stands out as the epitome of a daring climb. The chapter describing the achievements of the best Scottish mountaineers across a century was, on the whole, calmly delivered. But I sensed a note of awe in the author's summing up of Wild Country as a climbing achievement, that climbers' athleticism had become so impressive that it was now possible to tackle such a feature. I could tell this route represented something special, an opening up of new possibilities.

Reading this history of successive climbers taking on ever more extreme rock features on mountains like The Cobbler was the lens through which I discovered these mountains lying an hour along the train line from my house. As I read, I began to think about visiting The Cobbler on the train and exploring it. But after reading this, I didn't just want to see the mountain. I was already hooked on the wonder of these hard climbs, and I wanted to see them in the flesh.

Back at school, a couple of boys in the playground had started quizzing me about what I was doing in the hills. They'd seen me in the local woods near my house, trying to

Opposite: Peter McGowan in Arrochar on our first visit to The Cobbler in 1994.

learn how to abseil by wrapping a rope around my waist and thighs to create friction. I'd gained enough skill at climbing trees and my rope was long enough that I'd set up a couple of big swings that were more impressive than those in situ in the woods. Before this, I'd felt, at times, almost invisible at school. I liked it this way because the alternative thus far had been much worse. Attracting attention usually resulted in humiliation of some sort if things went well, or occasionally taking a few knocks and kicks if not. I had no physical confidence, nor the ability to hide this, which made me both a target and a sitting duck for other boys who wished to either demonstrate their confidence, or just entertain themselves.

So when some of them started to ask about the hills, I said as little as possible. *The quicker they lose interest, the less painful this will be*, I thought. But one boy didn't lose interest. As the others drifted away, a big lad I'd never spoken to before stopped me in my tracks by asking if I'd climbed The Cobbler. I wondered how he knew about it, and his very specific question disarmed me a little. I already knew he was called Peter McGowan, since he was one of the biggest and strongest boys in my year and regularly won the school cross-country race, a guaranteed source of social credit at school. At least half the cohort of kids wasn't fit enough to run continuously for three miles on grass, so the hard reality of being lapped by those who could was an unequivocal statement of their ability and your lack of it.

Peter's parents had a caravan on the Argyll coast, and so the trip up Loch Lomondside and past The Cobbler was a regular weekend adventure for him. Unlike me, he'd seen the unmistakable spiky skyline, and it had obviously left an impression on him, too. I felt uncomfortable being asked about hills and climbing as if I had some expertise, and I worried his questioning was just a more elaborate joke at my expense than I was used to. But then Peter stunned me by asking if we could climb The Cobbler together.

We agreed to meet on a Saturday morning and take the train to Arrochar. It was a sunny, breezy morning and we strolled round the head of Loch Long to the foot of The Cobbler. I felt excited but also slightly detached, as if too many new things were happening at once for me to absorb. Peter, on the other hand, talked straightforwardly, asking how long I thought it would take to get to the top, which of the peaks were difficult and what I'd packed for the day. He spoke to me as an equal, so I started to respond and share my excitement for being there and seeing all the hills I'd read about. Equipped with knowledge gleaned from *Scotland's Mountains*, I dare say it must have seemed as if I did know a lot about mountains and climbing. I had also carefully read and memorised how to build and equalise anchors using either natural rock features

Opposite: The path to The Cobbler, now redirected, used to follow the old Puggy line climbing directly up the slope from the head of Loch Long. Spotting other hillwalkers above us and trying to overtake them before the top of the Puggy line was excellent training. © Peter McGowan

or climbing equipment, and various ways to abseil. As we trotted along the path, I shared with Peter the history contained in this one mountain, from the early pioneers to the cutting edge.

Halfway up the approach path we came across two huge glacial erratic boulders, one with a small overhanging recess which had obviously been used as a shelter by hillgoers for a long time. We stopped here, resting, drinking from the river and talking. I started to try and climb the boulders. They were the first I'd seen other than those at Dumbarton Rock, and I was struck by how completely different they were in character, with rippled small edges and pockets. Within a few minutes, we'd climbed both boulders and excitedly looked around for harder lines on them. Most were too hard for either of us and we jumped, wide-eyed, from highpoints into the bog below. Our respective efforts seemed to hang on applying multiple different tactics to gain height on the rippled walls. Where one of us tried to use aggression and burl, the other got to the same place with an intricate sequence and a cautious approach. This playing field was remarkably level. Competition between us seemed to blend neatly into collaboration. The attempts were momentarily serious and worrying, but mostly lighthearted and unstructured. When the lines had us both stumped, we just sat in the grass eating multicoloured industrial cakes from the Arrochar shop. It was the first time I had experienced real enjoyment in sharing a sporting activity with anyone else.

Afterwards we picked up our bags and continued up the path. As it snaked below the cliffs of the North Peak, we craned our necks to inspect the overhangs looming above. The deep cracks and chimney routes of the Victorian pioneers looked conceivable. Further up, we traversed below the main overhanging beaks of rock that gave the mountain its name, appearing like a cobbler, hunched and leaning over his shoe stand. The left-hand overhang, 50 metres high and leaning out over the path, looked incredible to me. 'That is Wild Country, the hardest route on The Cobbler,' I told Peter. It was a day filled with firsts for me. I'd gained independence, new experience and a best friend. But the memory of seeing that smooth, scooped overhang held my attention long after my first day in the hills with Peter. I suppose I had been primed to focus on it from that note of awe I'd sensed in the description of the first ascent of Wild Country in *Scotland's Mountains*. Possibly I just shared the author's sense of wonder that a human being could have the skill, strength and confidence to go up that wall.

More importantly, like Dumbarton Rock, it was the concept that this monument to world-class rock climbing existed in my own back yard that I connected with. Something about this grabbed my attention more than the grandest of sporting spectacles on TV ever could. The overhang on Wild Country looked just like those

Left: On The Cobbler for the first time. Reaching the summit requires a short scramble.
© Peter McGowan

on the boulders we had climbed on the approach path, with the same rippled quartz edges. I had been amazed at how the puzzle of moving between the edges on the boulders could be solved in multiple different ways. Neither climbers, nor the famous athletes on TV, could shortcut the process of refining the strength and confidence needed for world-class performance. Olympic athletes had to challenge themselves a million times in training, in return for one chance to stand in an Olympic final. But the challenge of solving moves in climbing was different. Climbers could stand at the start line of the hardest routes any time. Nothing was stopping me from just asking the question, *How would you move from that hold to the next?* In a sense, I could experience a little bit of a cutting-edge climb every time I engaged with that question. Once asked, I felt irreversibly pulled in by the curiosity.

'Come on, Davey!' As I sat in the evening sun staring at Wild Country, Peter reminded me that we should get moving, or we would miss the last train to Glasgow.

05.

NO GOING BACK

Left: Peter McGowan on The Cobbler, with the summit behind, after climbing MacLay's Crack (III,3) in December 1994.

In the months that followed, Peter and I spent more and more time at school talking about climbing. In previous years, I had been one of those boys Peter would have lapped during school cross country. His experience of running at a club had given him an awareness of the need to train in order to get anywhere with sports. I found his willingness to do this in a self-directed manner without any input from adults almost perplexing: his initiative just wasn't something I'd come across among other kids. We'd go out into the woods after school, climbing trees, practising abseiling and making rope swings. But afterwards, Peter would point out that to be faster in the hills, we should run up hills. We started to run circuits around the streets at the end of the evening, sometimes adding short sprints up the hill to Peter's house. Keeping up with him was impossible at first. In PE classes at school I would previously not even have tried, but Peter's ready acceptance of my lack of fitness and our shared motivation to be better prepared for our mountain adventures turned my attitude around.

Peter helped me to abandon a culturally entrenched mindset that physical prowess in youth and degeneration in nearly all the middle-aged adults around us was largely predetermined and instead grasp that it is mainly a function of physical training. For anyone familiar with sport and training, it may be hard to imagine that this wasn't obvious to me. It was my first big lesson in how limiting culture can be when it goes unchallenged in your world. My bias here was entirely unconscious; I didn't even realise I thought this way until my new life forced me to reflect on it. It is deeply ironic that I reached my mid-teens believing that I would always be weak and poor at sports (and much else). My mum and nan had agency in spades. It was down to their grit and hard work that I now found myself running for fun in suburban streets and being influenced by kids who could show me a different way. Yet somehow, I needed another influence, and more time, to step outside this way of thinking.

The default culture of the west of Scotland that I knew was to assume that performance would be poor, that life would be shit. There were of course exceptional people who showed strength of sinew or character and reaped the rewards of possessing those qualities. It wasn't that my mum's and nan's formidable strength of character went unnoticed. It was my rationalisation of their abilities that was the problem. They were seen as exceptions - different from the rest of us. I feel that this attitude can arise as a coping mechanism to the rigours of life in a city as deprived and depressed as much of Glasgow is. The physical impact of hard graft and relentless stress, sometimes compounded by self-inflicted damage from alcohol and poor diet, all conspire to frustrate and eventually erode both ambition and ability. Someone living this life may feel powerless to change the picture, but they can at least change how they feel about it by saying, 'Whit's fur ye will no go by ye.'

I did not yet know the meaning of hard graft or stress, but I did soak up its effects in the culture around me. My runs with Peter helped me see that this fatalistic way of thinking was utterly flawed, and reflecting on it from the outside was more exhilarating than the running itself. Now I began to feel an exciting, if uneasy feeling, that it was not my mum, nan and Peter who were exceptional, but me. Of course, the strength of character they showed could be an important component of physical prowess in those who excelled in sport. But along with fitness, it was merely another outcome of training, not the cause of it. All that is needed to initiate this way of being is the desire for the outcome of training to run deeply enough. I wondered what other biases I might carry that I could target and retrain, just like the newly appearing muscles in my legs.

I marvelled at Peter's strength, athleticism and toughness. Pull-ups and push-ups seemed effortless to him, and his relaxed demeanour running uphill seemed almost unreal. Whenever I expressed my admiration at his strength, he quickly corrected me. 'Oh no. I'm not strong. My dad is strong.' Before I'd seen this for myself, I didn't really believe him. His dad is a traditional memorial stonemason, a six-foot-six bear of a man with huge forearms and the hands of someone who has wielded heavy tools for a lifetime. Later, as we left school and Peter joined his dad's family business, I worked with them over a summer school holiday. I watched Michael mason a huge block of granite with a hammer and chisel twice as big as I could imagine would exist. They looked like comedy tools from a film set, until I picked them up and felt their weight. Shards of granite splintered off in all directions as Michael effortlessly swung the hammer. In minutes, the unsuspecting lump of granite had become a headstone.

My own dad, when we met, would often greet me with fists up and a slow 'thunk' on the chin at the end of his jab, followed immediately by a big smile. His fist was applied

Right: With my dad, Norman MacLeod. He was in his late 50s when I was born.

just a little harder than it needed to be for a greeting. But I understood that this was just a show, not a real indication of strength. I'd still never really seen an 'in the flesh' demonstration of the physical strength that men can have. No doubt I would have learned about this from my dad, had things been different in our family. I just didn't get to spend time with him in the right context to see it. You can only have a powerful physical presence if you are actually there.

My library books were a consistent presence though, and I continued to immerse myself in them. I soon picked up another deep-seated thread of Scottish mountain culture, which was that winter climbing on snow and ice was considered the most revered discipline. Our stormy winter climate dictated long days, severe conditions, serious climbing and a commitment to grabbing opportunities on chosen objectives whenever they arose. Peter and I agreed we wanted to experience this type of climbing.

As the winter got going, I had a few days out on my own in the snowy gullies of the Luss Hills and Ben Lomond, getting used to kicking steps in hard snow and using an ice tool for balance. It wasn't a proper ice tool - I had found a roofer's hammer in my mum's loft and asked her if I could file a few teeth in the end of the pick. It was reasonably effective on hard-packed névé, and I gained some confidence moving on snow. Its primary utility, however, was to serve as a reminder of how serious it would be to take a slip on steeper terrain. I had taken a couple of small slides on easy-angled terrain early on. But when I'd tried to brake with the tool, the fat pick would not cut through snow and it immediately ripped out of my hands. Later, I couldn't believe my luck when a friend of my mum's visited and gave me a proper mountaineering ice axe. She had been given it by a partner but had soon decided that winter mountaineering wasn't for her. I was already equipped with a keen sense of care for my balance from moment to moment on steep ground. I now had a great opportunity to combine this with my new tool and get onto more difficult terrain.

In December, Peter and I returned to The Cobbler, which was plastered white with snow and rime. Winter climbs in Scotland fall into three broad categories. First, the easy gullies of steep snow, which in good conditions can be climbed with just a mountaineering ice axe for balance and as a brake to arrest a slip. Beyond this, harder climbs tackle either cascades of water ice running down the cliffs, or so-called 'mixed' climbs on snow-covered rock. These two classes of climbs, at least in modern times, are climbed with two specialised ice tools, which are shorter, with steeply downturned picks to hook more effectively on ice or rock flakes. Crampons with a protruding front point are also essential for any traction on pure water ice or verglassed rock.

Although this approach was the norm, I was aware that some of the easier mixed climbs, pioneering in their day, had been first climbed with one axe and nailed boots, before dropped picks and front-point crampons had been developed. Nailed boots were a forgotten technology in mountaineering, superseded by strap-on and then step-in crampons, which were far more effective and could be taken on and off depending on the terrain. I hadn't appreciated that nailed boots could partially fulfil the role of crampons and give some traction on icy cliffs. I'd seen photos of climbers in the early 20th century using one axe on mixed climbs, so I felt I had enough equipment to attempt The Cobbler's cliffs in winter, armed with just the rubber soles of my hillwalking boots.

Our first outing was a long mixed route on The Cobbler's North Peak called MacLay's Crack. North facing and not that steep, the cliff was festooned with turfy ledges and hairy moss, with blobs of blue and green ice dribbling from deep cracks. In summer it would likely be pretty unpleasant to climb. But frozen solid, saturated turf becomes a huge asset to the climber. When swinging the pick of an ice axe into it, its consistency is like ice cream straight out of the freezer. It feels surprisingly secure to

Right: Attempting the crux pitch of North Wall Traverse (IV,4) on The Cobbler using dropped pick ice axes and crampons for the first time. The crucial patch of turf on the crux traverse can be seen on the right wall of the corner.
© Peter McGowan

Opposite: Approaching the South Peak of The Cobbler in excellent winter condition. North Wall Traverse starts on the snow ledge at the lower left of the face, following the corner and ramp line leading rightwards to the sharp upper arête and the summit.

climb - at least with sharp points attached to all four limbs.

I only had the one, though. With my right hand, I grasped at the frozen turf, digging fingers into the back of the little tufts of grass that topped every rail or ledge. I carefully stood on the blobs with my feet, making the most of the friction from crusty snow that sat on top. After climbing the first third of the route, this seemed to be going well, and I was delighted to be making such steady progress.

But then a rightwards traverse to gain a big chimney system stopped me in my tracks. Glassy water ice drooled from a crack, covering the rock wall below. I stretched right with a leg and poked at the bubbles of ice with the edge of my boot. Obviously I couldn't stand with more than a small proportion of my weight on those bubbles without my foot skating off. I'd need to commit with both feet on ice for a couple of moves before reaching the chimney. Now, the need for a second ice axe filled my mind. There was plenty of frozen turf, but while I moved the axe, I'd be entirely dependent on a firm grasp of frozen grass and keeping my feet very still on the icy bubbles. Although I was a novice on this terrain, I could already draw on some experience of judging how much force I could apply to smooth glassy footholds of Dumbarton Rock basalt.

Delicately, I moved past the ice flow, acutely aware of the silence between held breaths, and reached the dark sanctuary of the chimney. The only equipment we had at the time were two karabiners I'd bought for making an abseil friction brake and two straps Peter had found in his garage, which we tied into slings to wrap around chockstones in cracks or drape over spikes. I'd used both our slings, so I spent a while trying any way I could to wedge the ice axe into the cracks at the back of the chimney to make an anchor so that I could bring Peter up. Peter had got two dropped pick ice axes and a pair of crampons for Christmas, and as I watched him tackle the icy traverse below, it was obvious that we'd be able to climb something harder with the right tools.

At the top of the climb, I jammed the axe behind a boulder and brought Peter up as the sun set. I was warm from the effort of thrutching up the chimney, but my face felt raw from the biting cold wind. My body and everything around me glowed; ice-encrusted mountains loomed under the deep red sky to the west, rich blue and purple to the east. It amazed me that one climb could yield such a spectrum of emotions. In the hours that preceded, I'd felt an intense drive to push forward, tolerating fear and dissatisfaction with temporary failure to make upward progress. Yet here, lost in the spectacular light display, none of this appeared to matter anymore, and I felt intense contentment just to be. Possibly these seemingly disconnected feelings were, in fact, interdependent.

Despite my contentment, I looked across at The Cobbler's South Peak. Steeper, north-facing and dark, it contained the hardest winter routes on the mountain. The easiest line on the face was North Wall Traverse at grade IV. One grade harder than MacLay's Crack. As Peter joined me to enjoy the fiery sunset across the Arrochar Alps, we wondered if we could lift our sights to this route.

A friend of Peter's had joined us on a couple of days out in the hills the previous autumn. Gary had done less than us but seemed keen, and he had plastic mountaineering boots, crampons and two ice tools. While arranging our next climb, Peter told me that Gary was going on holiday for a couple of weeks and had offered to lend me his kit. I could hardly believe my luck and jumped at the chance, set on North Wall Traverse. Between the three of us, we now had a few more pieces of gear, some of which we'd got for Christmas, some we'd found while practising abseiling at the crags and some we'd been given by a sympathetic, perhaps concerned, staff member at a climbing shop in Glasgow who had asked what kit we had for our winter adventures.

Peter and I stomped in deep snow to the foot of The Cobbler's South Peak. It looked a big step up in difficulty. Large overhangs of schist loomed above, dribbling with ice and lacking the big cracks and chimneys of the North Peak. The crux of the climb was the first pitch, up a corner into the overhanging wall above, then out across the steep right wall to gain a huge ramp of frozen turf that led more easily beyond. Intimidated, I gingerly moved up the corner. The sense of newness was overwhelming as I tried to get used to hooking and torquing the dropped pick tools in the corner crack, edging on ripples of schist with crampons on my feet for the first time and trying to shake off the general atmosphere of the gloomy north face. It seemed to whisper, 'You are moving well out of your depth.' But I had to give Gary his crampons back after the next weekend and there was no way I was backing off without giving this everything I had.

I reached the point where the corner crack fizzled out into the overhangs, and I must now move across the smooth right wall. A long moustache of frozen turf led the way, and I was exhilarated to swing across it, planting the ice tools securely while Gary's front points bit tiny ripples of quartz on the vertical wall. The big ramp beckoned just a body length away, but I'd need to stand on the turf I was hanging from to reach it. There was a rounded crack in the wall above, and I hooked the pick of Gary's ice axe into a tiny slot in this. The hook was thin, and the pick dropped a centimetre into the crack and no more. Shaking, mouth dry with fear and feeling like I was asking for trouble, I pulled up on the hook. I'd need to make the committing step of removing the other ice tool from the relative security of the turf below and take one more hook in the wall to reach the ramp.

I couldn't dare commit. It felt too extreme, with an impending sense that everything was about to go wrong. Yet that ramp was just a couple of feet away, tempting me on. With arms tiring, I moved up and back down several times, each time progressively less in control until it was clear I would soon fall off. The strategy of hesitation was about to backfire, and I shouted down the inevitable decision to Peter that I was going to reverse.

With every move back down the corner came increasing relief. But on the final step into the deep snow at the base, the relief was replaced by a sense of anti-climax. I didn't know what to do with myself. I stood looking at Gary's ice tools and crampons and the adrenaline drained away. Peter was shivering; we should move. We packed away the rope and I stopped and stared at the gloomy walls as Peter retraced our posthole steps back down the corrie snow. 'Come on, Davey!' he shouted after me, by now familiar with my habit of staring endlessly at climbs I'd just failed on. As I stomped in the postholes behind him, I became overwhelmed with an emotion stronger than any I could recall feeling thus far in my life. Stronger and more unpleasant than the fear I'd felt a short time before. I had wasted a huge opportunity. As I walked away from the wall behind me, it felt like it was burning a hole in my back. I couldn't let go of the image of the crack at my high point. There was another feature to hook. I'd seen it; I just didn't dare pull up to it. Why not? Only intimidation and nothing else.

The feeling churned in my mind every day at school the following week. What was stopping me from just pulling up on that hook and reaching one move higher? Obviously I was scared of falling, and the intimidating nature of the steep, icy cliffs amplified that. But the hook was good. The unavoidable reality was that I'd wasted my chance, and I hated it. Even worse, I still had Gary's ice tools for one more weekend, but I'd have to stare at them in my room, since Peter was away then and couldn't come for a rematch.

The forecast for that weekend was excellent, so I decided to go to The Cobbler on my own and climb something easier to make the most of Gary's tools. I didn't decide what. It annoyed me too much to even look at the guidebook and flick past the page with North Wall Traverse. I remembered there was one grade III that shared the same start as North Wall Traverse but moved off left and around the corner onto the straightforward south ridge of the mountain. I'd climb that solo.

Arriving below the face, it still frustrated me to look at North Wall Traverse, and I approached the base with my mind still rather stuck in the previous weekend's events. I moved up the initial few metres of steep snow and turf to where the routes split and

Opposite: On The Cobbler with the South Peak behind. The diagonal ramp of North Wall Traverse cuts across the face from left to right.

stared up at the corner and right wall where I had failed before. It still looked just as intimidating. Without really thinking, I moved up into the corner. I could still remember exactly where I had hooked Gary's tools, but without the shock of the new, each move now felt much easier and less scary. Seemingly in a few seconds, I had climbed several metres up the corner. But where I'd felt scared and out of my depth before, I was now relaxed, despite having no rope. Enjoying the feeling, I resolved to make one more move up, then go back down. But many more 'one more moves' further I was moving out across the wall, placing the front points precisely on the familiar quartz ripples. The curiosity was too much now. Was that hook really as good as I thought? I reached up and, with a soft 'thunk', seated Gary's axe in it. It was solid. But the situation still felt beyond my comfort zone, beyond me. Just like before, I started to shake with fear, and I gingerly brought the tool back down to the turf.

The situation had become very simple and clear. My mind was in a straightforward deadlock between curiosity and fear of what the move would be like. But rather than viewing this as a fork in the road at which I could either go one way or the other, it felt like I was just spinning in a circle. No decision was being made, and all the while the strain in my arms was growing. As the pressure to decide reached an almost unbearable level, another part of my mind noticed something happening. My left arm was removing Gary's ice tool from the security of the turf. It was immediately replaced by a foot swung high and planted sideways to stab in all 12 crampon points. As my hips opened and my body leant inwards and upwards, I drew cold air deep into my lungs on sharp breaths, anticipating the outcome. The left tool found a second hook. There was one, after all! Stood up on the turf, the angle eased slightly, and with eyes bulging, I reached up without hesitation and whacked Gary's tool into the lowest tufts of the ramp. As my tools sank into the solid placements, I could feel the hairs on my neck relax and the jitter on my breath subside.

Moving up the ramp, I paused to take in the position. Above, the overhangs looked spectacular and dramatic, but the ramp cut through them, and I continued in a rhythm of movement that felt easier and easier. Now on steep snow and secure frozen turf, I floated on a wave of euphoria. The 'out there' sense of exposure above the corrie I could no longer retreat to carried no weight in my mind. If anything, the sense of upward momentum from one solid placement to the next felt like a form of gravity in itself.

Sitting in the sun at the top of The Cobbler's South Peak, I had an intense feeling that I could not explain or understand at the time, and which only later came into focus. I was not the same person I had been at the foot of the wall, an hour or so before.

For better or worse, the curiosity that had pushed me through that move on the wall was a committing step, and there could be no going back. It was beyond doubt that my ability to climb was not fixed. It was, in fact, entirely fluid. As I looked across the corrie, the huge overhang of Wild Country glinted in the sun. That awe I had noted while reading about it in *Scotland's Mountains* and then sensed myself on my first visit to The Cobbler was still there, but had also completely changed. It was no longer that I couldn't grasp how someone dared to consider climbing such a feature. Now I just looked at it in awe because it demonstrated how this leap of confidence could be made. The climb I'd just soloed had helped me to turn my perspective on its head.

06.

NO RECESS

Left: Dumby scenes.
© Claire MacLeod

I didn't see it coming, but the intensity of winter climbing in the mountains upturned my experience of the real world by transforming how others treated me. On Mondays at school, my mind remained lost in a neurochemical glow that still burned from the weekend's adventures, replacing any pressure to find my place among peers, with all the problems that brought. As far as I knew, this glow was invisible from the outside. I hoped it would make me even more invisible, but much to my confusion, it had the opposite effect. As I further disengaged with school, it began to engage with me.

The first time I noticed it was in a circuit training class in PE. Almost all mandatory school activities, except fighting, didn't really use upper body strength, so my improving arm strength had flown under the radar. I lined up at the bar alongside two other boys and knocked out 12 pull-ups, still finishing my set well after they had slowed and dropped off. Boys soon asked me more about climbing, and this didn't surprise me, since physical prowess was a measure they kept score of. What did surprise me was that they began to talk to me rather than down to me. I interpreted the change in their behaviour as being solely related to my improvements in physical strength, and this made the attention feel silly and fake. Later, I wondered if they read a far more subtle change in my confidence that I was not yet aware of myself. Regardless of the reason, it did make the school days pass easier, and humiliation was far less frequent.

With Peter's influence, I started to try in the school cross country run. My hill days meant it wasn't difficult to make the jump from being one of the kids who'd be dawdling round, wishing I was somewhere else, to finishing near the front and looking forward to the runs. I was encouraged by the PE teachers to enter some cross country races against other Scottish schools, but the events were never much fun, with a lot of time spent just waiting around.

Opposite:
Feeling happy at Dumbarton, the Requiem face catching the sun behind, always difficult to ignore.
© Claire MacLeod

The Scottish Schools Cross Country Championships were considered quite a big deal, and we were urged to put in our best effort. Peter was running really well and expected to get a high placing, but when the starting gun went off, the runner in front of him in the crowd rocked back, and his racing spikes punctured Peter's foot. I don't know if this was an accident, but I always wondered if perhaps it wasn't. At each bottleneck on the course, boys shoved and elbowed to get ahead. At the finish line, we were funnelled into a single file so placings could be recorded, and one boy rammed past me. I stood for a moment in disbelief at his lack of sportsmanship, until another two boys shoved me over. The race was a huge eye-opener for me.

After the race, the PE teacher asked Peter what had gone wrong. He was pretty unsympathetic, even when Peter showed him the bleeding holes in his foot. I was in no hurry to participate again. When the next Saturday inter-schools race came around, I went climbing instead. The following week at school, my PE teacher asked why I hadn't shown up. I didn't have the confidence to articulate my distaste for the ethos of these events and offered a superficial explanation of wanting to go climbing. He correctly pointed out that I was letting a team down and should have come because that's what I'd said I would do, and he gave me a punishment writing exercise to take home. As far as I was concerned, I was writing out the death warrant of my mainstream sports involvement. I associated it with rigidity, pressure to perform, cheating and social credit for all the wrong reasons, and I didn't want any more part in it.

The following year I was preparing for Highers exams, but I had become more and more disillusioned with school. I wasn't afraid of learning or schoolwork, but the more I stayed away, the harder it was to go back. In its place, I spent more time at Dumbarton Rock, working my way through the boulder problems. While the school environment pushed me away with every staged, fake interaction, the climbing world pulled me in. Like any kid, up to this point in my life, I took the world as I found it and accepted without question the nature of others, and myself. My peers had the confidence to go out and form or abandon relationships as they wished. While this seemed impossible to me, they made it look easy, though I may have been far from alone in concealing my difficulties. My inability to integrate well led to anxiety; the social world felt unpredictable and I needed to be hypervigilant too often. Not understanding why I felt different added a layer of confusion which only deepened when suddenly I started to fit in at school. I was the same person, or at least felt that way. As far as I was concerned, the only thing that changed was that I could now do 12 pull-ups and run fast. Was that really the sum of my value?

As I matured, that confusion gradually morphed into a dark tendency towards self destruction. What appeared to be the only thing that gave me status amongst my peers was what mattered least to me: my place on a scoreboard. This hollow victory made me feel worthless, and my downward trajectory rendered school attendance unimportant. Climbing had deflected this fall, which was on the verge of getting out of control, and bounced me in a different direction.

I was called in repeatedly to explain my absence to my perplexed head teacher, who, understandably, had no choice but to refer me to the Education Officer from the local council. I sympathised with his position. So I took a train to the council offices in Dumbarton - the same train ride as my regular commute to Dumbarton Rock - where the attendance officer sat me down in his office and politely asked me to explain why I'd been sent. I told him I felt very uncomfortable around the other kids at school and hated the fighting and humiliation. In class, the teachers often simply read aloud from a textbook, and I could do this on my own, after dark, and use the daylight to climb instead.

After I'd finished, the officer sat back in his chair, pausing for a long time. I fully expected to receive some sort of punishment as well as a simple demand to get back to full attendance, but his response wasn't what I expected. He began by thanking me for giving a clear and honest answer and told me he did a bit of rock climbing himself and understood what I was talking about. He then continued, 'My advice is to go to school as much as you need to keep yourself out of trouble. Once you get to university, I think you'll be fine.' His approach was so different from my head teacher, who didn't care why I wasn't coming to school. The attendance officer had listened intently to my reasoning, occasionally reflecting and asking if he understood me correctly. I had rarely experienced this type of interaction with teachers, and something about it shifted my attitude toward school. I took his advice, and attended more often in my final year, but my total focus on climbing cost me in exam results that year.

The following winter, Peter turned 17 and got his driving license. He could use his mum's car at weekends and our winter climbing horizons suddenly opened up well beyond day trips to Arrochar on the train. Our obvious targets were the bigger mountains around Scotland - Glen Coe and Ben Nevis. The prospect of winter routes here intimidated us; we knew these ranges were not to be taken lightly, and the lack of some key pieces of equipment - my own ice tools, maps, head torch, etc. - was still a major limitation with few workarounds I could think of.

Ordnance Survey maps were expensive, and I couldn't afford to buy one for every new climbing area I visited. Instead, I'd spend ages in the outdoor shops looking at the maps and memorising the layout of the hills we planned to climb. Peter had often remarked that I appeared to be able to follow my nose in a blizzard and find the right way. The truth was that I recognised buttresses, the fall line of slopes and distances between features from poring endlessly over maps in the Nevisport shop on Sauchiehall Street in Glasgow.

It was stuffy in the shop, and on one visit I had my green fleece jacket half unzipped as I tried to memorise the layout of Ben Nevis. As I folded the map away and put it back in its place, a staff member made eye contact with me. I dare say I looked awkward, feeling guilty for staring at the maps for the previous hour. I milled about a bit downstairs, looking at climbing gear, and had an odd feeling that a tall man with a moustache was taking the same path as me through the store. As I headed out the front door, he suddenly appeared at my back and crashed into me. For a second, I assumed he'd tripped on the step and let out a reflex, 'Oh, sorry', then yelped as my arm was twisted up my back and my face shoved violently into the glass shop front. Once he'd got a firm grip of my arm, I was marched back into the store up the stairs, grimacing from the pain, as all the staff stood looked on, stony-faced. I was led into a small room and told to sit, with staff and Mr Moustache crowded over me.

'Right, where is it?' said the woman who'd made eye contact with me.

After a bit of 'Where is what?' she said angrily, 'The MAP! I saw you put it in your jacket.'

'No, you didn't,' I responded more firmly than I expected, given that I was terrified and my hands were trembling. 'It was map number 41, and it's still on the shelf downstairs, where I left it.'

Mr Moustache interjected. 'Well, if you're not going to hand it over, we'll search you for it.'

I said nothing and raised my arms. After a search of my person that was unnecessarily thorough for a five-inch map, the staff stood back and looked at each other. The woman returned from checking the map display on the landing, looking rather mortified.

'We're very sorry; I was sure I saw you put it in your jacket.'

Still shaken and sore, I didn't say much apart from asking if I could go. They shuffled away from the door and I walked out, trying not to cry. When my mum rang the store later, she was shaking too, but not in the same way as me. I could barely listen to the dressing down she gave them over the phone.

A couple of weeks later, the store manager phoned my mum and said he wanted to see me in the store again and he'd like to give me something from the shop. I wasn't

Opposite: On Beinn Udlaidh with the Cassin ice tools given as redress after the Nevisport incident.

sure, but the chance to get some more climbing gear was too good to pass up. Ice tools were top of the list, and it happened that a pair of Cassin axes I really liked the look of were on an end-of-line discount. I was happy to accept them as a peace offering and, more importantly, the opportunity they gave me to throw myself at some harder winter routes.

My new tools helped me rapidly gain confidence and speed on steep ice. But as Peter and I ventured onto Ben Nevis, I soon learned that this was still not enough. We weren't very fast at moving through pitches, setting up belays and generally avoiding faffs, nor were we practised in retreating if things weren't right. We didn't really appreciate how or when to start abseiling back down a big ice route where anchors were hard to find. Being caught out in darkness high on a climb certainly still scared me, not least because I didn't yet have a head torch.

Another feature of Ben Nevis climbing that we wanted to stay well away from was avalanche danger. Standing under the north face on a blowy day with heavy snow falling, gazing at the slopes leading up into Coire na Ciste and Observatory Gully, we didn't really know how to judge whether they were safe. We did have enough good sense to assume they weren't, and chose a Grade V icy mixed route called Pinnacle Buttress on the west face of Tower Ridge, which could be accessed without moving across the big snow gullies.

With Peter stood in a flat bay of fresh powder that was getting deeper by the second, I set off leading the first pitch. Waves of snow and spindrift poured down the face, making upward progress arduous and intense. But this was also what I loved about winter climbing. Seeing huge volumes of snow deposited and piling up on mountains still filled me with joy. After 50 metres, Peter shouted that the rope was running out and I'd have to find an anchor. With the rock obscured by ice and snow, finding cracks was desperate, and I became self-conscious about my lack of progress. Finally, my new ice tool found a thin vertical crack. Peter had bought a peg recently, and as I started to hammer it in, the snow caking the surface of the crack began falling into it. The next moment, I realised with horror that the crack I was hammering the peg into was actually a detached block, taller than me, that was now peeling off the wall.

The block was gently leaning against me, just a fraction past its point of balance. I tried to push it back, but it wasn't going to happen. I knew if it leaned out another centimetre, I'd not be able to hold it back, so I screamed down to Peter that a big block was coming down and to get in against the wall. With confirmation that he was out of the way, I stepped to one side and the fridge freezer-sized block sailed through the air, belly-flopping into the snow, two feet in front of Peter, with a huge boom and a powder snow explosion.

Peter leapt into the air in fright. Understandably, he wasn't too happy, and I apologised and built the anchor in the freshly exposed dry rock. Five pitches of steep ice followed without incident, but with each pitch, I became increasingly aware that we had been on the face a long time. With momentary relief, we popped out onto Tower Ridge right below the iced-up bastion of the Great Tower itself. But a gathering gloom quickly darkened the mood, and the weaknesses in our climbing skills filled my mind once more. At that time, we both viewed being caught in the dark as a serious issue, but in reality, climbing beyond daylight hours in Scotland is normal and shouldn't be concerning once you're familiar with it. Many experienced winter climbers will still feel a tinge of urgency as the gloom of dusk encroaches while high on an icy cliff, especially if a blizzard is raging. But once it is fully dark, it's often easier to slow down and take each forward move one at a time. Your world is narrowed to that of your torch beam, which can somehow simplify the decision-making process as you continue. But first, you need a head torch.

We hastily moved round the Eastern Traverse and reached Tower Gap, the crux of Tower Ridge. Having already completed Pinnacle Face, our singular goal was now

Right: Peter McGowan outside the CIC Hut on Ben Nevis before climbing Pinnacle Buttress.

just to get down. So when I spotted a snow ledge leading off left into Tower Gully, I suggested we take that exit as it would be quicker than climbing another pitch across Tower Gap. That way, we could race up the much easier Tower Gully and get to the summit plateau with a few more precious minutes of light to make navigation easier and safer. Peter agreed, and in no time, we were approaching the final climbing obstacle of the day: a large cornice rim at the top of the gully. The change of plan had worked, and we would make the plateau before it was fully dark. Peter reached the cornice first, and immediately began to dig a tunnel through soft snow in its centre. Without saying anything, I started traversing left to see if I could sneak around the biggest part of the cornice.

After a few steps left, I heard Peter whoop. 'I'm through!' he shouted from his hole as he shoved his ice tool upwards into the cornice.

At that moment, the entire cornice rim collapsed.

I was leaning back slightly, looking at Peter, when the huge mass of compact snow tore me backwards. After a few seconds of confusion, I had time to consider the situation as I somersaulted down the gully we had just climbed. As I tumbled, immersed in what felt like a river of concrete, snow filled my eyes and mouth, and I couldn't tell which way was up. At first, I behaved as if I might have some sort of control over what happened next, desperately clawing at anything with my ice tools. At one point, I felt a little resistance and seemed to slow slightly. Then a horrible thought crossed my mind. Observatory Gully, directly below, is broken by a 300-foot barrier cliff called Tower Scoop, one of my first ice routes on Ben Nevis. The avalanche would soon pour right over this face, and if I was still in it, I was finished.

It quickly became clear that trying to stop myself while inside the avalanche was utterly futile. The effort of wrestling to find the bottom while smothered in tonnes of snow had already exhausted me. I accepted that there was no way back and gritted my teeth, waiting for the inevitable and hoping that it would be immediate.

The thundering freight train of snow accelerated, and then suddenly the immense weight of snow gently lifted away from my body as I shot off the edge of Tower Scoop. *Here it comes.* In the blackness, deep in the mass of falling debris, there was no air resistance and it felt as if I was floating along with the soft lumps of snow. Only a sickening feeling in my stomach signalled my downward trajectory. Down was coming up fast, and I braced for the impact from below.

Bang! The strike came, but from behind, rather than below. It winded me and threw me into a tumble again. But I was not dead. The soft avalanche debris piling up ahead of me as it surfed the icy face of Tower Scoop must have smoothed the angle change

Top left: Avalanche debris piled near the foot of Observatory Gully. Tower scoop can be seen partially in the cloud, left of the black overhanging buttress of Echo Wall.

Bottom left: At the top of Tower Scoop, Gary Taylor seconding. This would be the exit point where I would be carried down the route by the avalanche in Tower Gully above.
© Peter McGowan

to Observatory Gully. I tumbled onwards, my body now limp and exhausted, every muscle burning from fighting to stabilise myself. I felt it slow. Resistance from the slope below pushed upwards, and I saw a momentary flash of light. The deep rumble gave way to a soft and gentle hiss, and the huge mass of snow drew to a stop.

I was on my side, half my face protruding from the snow, and through one eye, I saw the moonlit outline of my ungloved hand sticking out of the snow. I got it to work and clawed at the snow. At first, I had no power in my free arm and I repeatedly paused, almost apathetic as I let it lie on the surface between bouts of digging. Both muscles and brain needed oxygen and as soon as I cleared the snow from my compressed thorax, I felt the energy return and the pace of digging accelerate. Cold water running across my chest and legs also sharpened my hypoxia-dulled mind. Snow had been driven into every gap in my clothing and it now melted on contact with my skin, hot from the effort of fighting for life. For a few moments, I sat on the edge of the hole, hyperventilating and groaning involuntarily with shock and fear. Blood dribbled from my mouth onto the pristine snow, and the sight of this snapped me back into a coherent thought pattern. Where is Peter?

I scanned the debris, which seemed to cover hundreds of square metres all around me. The avalanche had fanned out across a convex slope, likely the reason I had not been buried. I got up and stumbled around, looking for him. He could be anywhere in this, suffocating right at this moment. Where would I even start? I needed help, and avalanche probes. Then, a thought crossed my mind. How far away was the CIC Hut, the mountain hut at the foot of the north face cliffs? It would be full of climbers, and I knew there was a store of mountain rescue equipment there. That would vastly increase the probability of finding him in time. But where was I now? Observatory Gully is 1,000 feet of climbing and takes nearly an hour to plod up in deep snow.

The weather cleared slightly, and moonlight gently illuminated the cliffs through thin cloud. I looked up and tried to recognise the features in the gloom. Where were the walls of Tower Scoop, Indicator Wall and Gardyloo Buttress that surrounded the upper part of the gully? Tears filled my eyes as, utterly confused and helpless, I continued to stare, shouting out, 'Where the fuck am I?' Then the features started to make sense. *That looks like the bottom of Point Five Gully up there ... That looks like the bottom of Observatory Ridge up there ... Fucking hell, that's not the Great Tower, that's the Douglas Boulder above me!* I could hardly accept what my eyes were telling me. The features I was recognising were 800 feet lower than those I was expecting to see. I had fallen the entire 1,000 feet of the north face and had exited the bottom of Observatory Gully.

With that thought came another: The hut was just minutes away. I took off down the slope, leaping, sliding and falling as my body trembled in shock. At the hut door, I felt a wave of relief to see a friendly face and an immediate invitation into the hut. The inside of the CIC is always cosy, packed with mountaineers all winter long. The long kitchen table was rammed with climbers who set down their forks and spoons as I gasped out the story. I told them I hadn't seen my partner fall in the cornice collapse and that he could well be in the avalanche debris. I felt humbled when, without a word, the entire audience of 20 mountaineers rose from their seats in unison and sprang into action, immediately reaching for salopettes, boots and head torches. The man who'd answered the door picked up a large rescue satellite phone mounted on the wall, called the local mountain rescue team and informed them the hut crew were going out to search and would need assistance.

Within a few minutes everyone began to file out of the hut door into the night. Just then, the satellite phone rang. The same man answered and listened.

'Yes, we have Dave MacLeod here … Oh, that's great news!' he said and signed off. Peter had just rung the mountain rescue team to raise the alarm for me.

It transpired that when the cornice had collapsed on us, Peter had been shoved down into the hole he'd dug in the soft snow. He'd stood up and said, 'Oh well, that's that problem solved then, Dave,' only to look round in horror to see that I had disappeared with it. He had pulled over onto the summit plateau, and after a bit of wandering about in circles in the mist - I had the compass in my bag - he'd sprinted off down the main path towards Glen Nevis, where he came across a pair of climbers with a phone. They had called the rescue team, finding out immediately that I was safe.

The episode gave us both something to think about. Safety in mountaineering comes from bitter experience, and this was one of many frights that served to catalyse a review of our skills and to move to another level of competence. Learning to climb independently had been really good for me; it had helped me cultivate an analytical approach to climbing. Yet it had its downsides, leaving me with gaps in both knowledge and confidence that could have been filled in sooner by spending time with someone more experienced on bigger mountains like Ben Nevis. Nonetheless, I think I may have been more vulnerable to lapses in planning, improvisation and humility in my climbing, had I acquired a rounded knowledge of the mountains without a real scare as part of my learning.

Had we held our nerve on Tower Ridge for a few more minutes and not worried as much about the dark, the epic may have been avoided. That said, in different snow conditions, the quick escape up Tower Gully would have made good sense with faster progress to the summit plateau in the last light. Speed is often safety in mountaineering, and these days, I could quickly get myself off Tower Ridge at night in a storm via multiple routes.

To this day, if I go into Nevisport on Sauchiehall Street, I feel a slight tinge of anxiety, a vestige of my experience there, which at the time greatly unsettled me, a timid young boy desperate to avoid trouble. Falling the height of Scotland's biggest north face left no such mark. I was back on Ben Nevis the following weekend, getting on with the next route. Mountaineering is essentially a game of getting yourself into trouble and getting yourself back out of it again. Something about the nature of mountains as the platform for this game attracted me. Whereas conflict between people made me feel lost and tempted to withdraw, misadventures on mountains had the opposite effect. I was beginning to understand the limits of this game, and I found myself wanting to get closer to that edge.

Right: Peter McGowan on the summit of Ben Nevis.

07.

CRAIG'S WALL

Left: On the 1990 Traverse (Font 7B) at Dumbarton Rock.
© Claire MacLeod

Back at Dumbarton Rock, I worked my way through more and more of the boulder problems listed in the newly published and rather uninspiringly titled *Lowland Outcrops* guidebook. Spending a lot of time at the Rock, I was gradually exposed to more climbers, too. It was heartening that, almost without exception, they were friendly and spoke to me as an equal from the outset, despite knowing nothing about me or my background. Others working their way through the guidebook often complained that some climbs seemed much harder or scarier, even if they had the same grade. I had climbed 6a (British technical grade), and one older climber who was trying the problems graded 5b would often ask me how to do the moves. When I demonstrated them for him, he seemed very grateful. In passing, he mentioned that he was a doctor. I was taken aback that someone in such an authoritative position would talk to me as an equal, and it stuck with me for a long time. The contrast with school life was striking - the Rock felt like a welcoming place, drawing me in to spend much of my time there.

It was true that, at each grade, some of the problems were harder or more dangerous than others and had few ascents. Only the very best climbers would go after and actually manage those 'outliers', and I began to gravitate towards the idea that I'd love to do all of them, or at least get as close as possible. I reasoned that it would act like a catalyst to becoming a much more capable climber, broadening my ability to cope with different terrain and helping me get a good head for scary routes. In situations like those Peter and I had experienced on Ben Nevis, I was aware that under stress, I would make decisions that I'd later feel were hasty or incorrect.

I set a simple goal, actually more like a rule, of climbing one thing out of my comfort zone on every single visit to the Rock. This was one of the best decisions I made in my climbing apprenticeship. It was a small enough commitment that I never avoided doing it, and there were so many lines to try that I was never short of opportunities to push myself a little. It was a very gradual, incremental process. Importantly, the rule related to feeling out of my comfort zone, not to any external outcome like completing a climb or a grade. This helped to anchor the training stimulus at the perfect level over time. It saved me from going after climbs I wasn't ready for and overwhelming myself, thus destroying my confidence. It also stopped me from allowing myself to game the system I'd set up by restricting myself to 'easy' climbs for a grade that played to my strengths.

Some of the really high boulders were basically free solos, where the landing was bad enough or far enough away that falling was really not an option. At minimum, a broken ankle would be virtually guaranteed. I'd climb steadily upwards to my previous high point and just try my best to push on a little further, even one move. I felt no pressure to go higher on any particular attempt, so I'd often climb back down, rest and then try again. Even just lingering at the high point, tolerating the fear or discomfort for longer, rather than immediately trying to escape it, was enough. I often found that after a few minutes, the discomfort dissipated a little, and I had space in my mind to look carefully at the rock above and figure out a potential sequence for the next move. Then I'd inch up until I felt on the edge of control again, and reverse back down. I became pretty competent at recalling the exact sequence I'd used to get to a high point and could climb back down in control.

With this approach, I picked off the scarier boulder problems and easier trad routes at the Rock, one by one. When the next summer came around, I was able to get back out to other trad venues. I often stayed at Peter's caravan on the Argyll coast, and we'd cycle to a great crag called Creag nam Fitheach, where I decided to try and climb my first HVS (Hard Very Severe), the last grade before the extreme scale started at E1. My newly won composure from climbing up and down the boulders allowed me to relax enough on lead to arrange the protection properly and feel confident that a good placement would hold a fall. For the first time, I didn't feel like I was soloing and felt safe climbing up to my physical limit so long as I could find protection. In one day, I did my first HVS, E1, E2, E3 and E4. I could see that I had certain ingredients for the overall challenge of hard climbing, and I'd just added another critical one that made the whole thing work so much better. I was still a little too keen to keep that grade progression going and was desperate to climb my first E5, but the reality was that I needed more time to consolidate my overall game at E4 or even E2.

I did scrape my way up a handful more E4s, which were finished with a dry mouth and trembling with fear. My attitude was that climbing hard was bound to feel hard, and that part of it was just learning to absorb fear and push on. This is true, but I was

Right: On the first ascent of Dressed for Success (Font 7B+) at Dumbarton Rock, using a three-finger openhanded grip on the holds to avoid aggravating injured and taped fingers.
© Claire MacLeod

taking it way too far, and I had poor awareness of tactics and how they could vastly change the difficulty of a climb. Some trad routes were prone to getting dirty if they hadn't seen any traffic over winter or weren't very popular. There just weren't that many mid E-grade climbers, and routes could go years without an ascent. Often, it would take someone operating at a higher grade to make an ascent and clean it as they went, but climbers sometimes took it in turns to abseil down a route and clean it, chalking up the holds so that it was much easier for a partner to climb 'on sight' from the ground up. Word would get around that a climb was chalked up, and others would home in and do it while it was easier, but I was ignorant of this tactic at the time.

Right: Taking a big fall during an onsight attempt on Production Line (E6 6c) at Cambusbarron. © Ali Weir

On a humid evening at Dumbarton, I went to try an old neglected E4 in the alleyway leading around to the main cliff. It was north-facing with a big grassy slope above and was dripping wet all winter. The routes here had long been out of fashion and were generally overgrown and manky. Belayed by my girlfriend, Claire, I set off up a groove, placing a cam behind a good hold after 20 feet or so. Above, protection was much less obvious, and I had to fight to place some shoogly wires in a thin flake that seemed unlikely to hold a fall. I got really pumped doing this and should have reversed for a rest, but instead I attacked the blind wall above with tiring forearms and fingers rolling on holds that were caked in soft lichen with the texture of paste.

At first, still in control, I moved up on fingery undercuts, trying to reach over a bulge to a dirty, rounded ledge. The holds above were covered in wet moss, and each foray upwards became increasingly rushed and desperate. Finally it dawned on me that I was in serious trouble. I was too pumped to carry on but too pumped to climb back down. I could stay where I was for only a few more minutes, and I had an increasingly panicked discussion with Claire about my predicament. I could hear her crying below as I made one final attempt to push on. But it was no use, and I watched my fingers open and felt myself peel off the wall and plummet.

I flipped over in mid-air and recall seeing Dumbarton's tower blocks upside down and hearing my dubious wires pinging out of their placements, tinkling like a distant dinner bell. I was about to crater into the ground head-first from 50 feet. But Claire, who had been obliged many times to watch the recently released climbing movie *Hard Grit*, had deployed her own climbing tactics.

With plenty of time to anticipate my fall and a warning that my wires would likely not hold, Claire knew that my cam placement was below halfway to my high point. At the exact moment I parted company with the rock, she bolted backwards and dived across the alleyway. As the ground rushed at my face, the rope came tight and I swung in a big arc, the long summer grass gently brushing my hat off. I flipped the right way up, stood up and looked at Claire, lying in a heap in the mud on the other side of the alleyway. I picked up my hat from the grass, and at the sight of it in my hand, I started to laugh. Claire started to sob. I felt terrible, but I was extremely grateful for her decisive action. After this, I went right back to E2.

Soon afterwards, I had to step back much further. Still ignorant of the importance of good conditions for climbing, I was frustrated by my inconsistent progress on the Dumbarton boulders. Usually I had a few projects on the go at once, and I'd try one for an hour or two then move on to the next when I was losing strength for high-quality attempts. I had already climbed a couple of the British 6cs at Dumbarton, and there was only one climb graded 7a (Font 7B+). It was called The Shield and had recently been put up by Malcolm Smith from Dunbar, who had shot to fame by repeating some of the hardest sport climbs in the world at the time. The Shield was only a couple of moves long, and I could actually do the first move, pull up and lunge for the top hold, a smooth sloping edge with no friction. Because I could touch the final hold, I felt like I should keep trying, though in reality I was quite far from actually doing it. I tried it over and over again in hot, humid conditions. The smooth, fine-grained basalt felt slippery and greasy, but I attributed this to a lack of strength.

In frustration, I kept attempting The Shield until I couldn't hold on any more. Then, one day, I began to feel a sharp pain in one of my fingers every time I pulled on the small crimps. Almost nauseous with the pain, I knew something was wrong. After a few days' rest, I tested the finger by pulling on a door frame, and it was no better, so I

Left: Climbing Mugsy Traverse (Font 7B) at Dumbarton Rock. This climb was my project when I had to take a four-month layoff from climbing with a severe finger pulley injury. I was shocked to climb it easily on my first day back to climbing after the layoff.
© Claire MacLeod

made an appointment to see my GP, who told me to take six weeks off climbing, take Ibuprofen daily and then come back to see him. Six weeks is a long time for a teenager, and I practically watched the clock before my follow-up appointment. With dismay, I explained to the doctor that, despite total rest and Ibuprofen, the finger was just as painful when loaded. I hung on his every word, assuming that whatever treatment he offered would be the very best available. He simply told me to try a different sport.

As I walked down the street away from the doctor's office, I had a huge shift in perception. Until that point, I viewed someone like a doctor, or in fact anyone in an authoritative position, with absolute trust and esteem. Though his advice was given earnestly, I simply couldn't accept that giving up the activity I loved, just like that, at age 16, was the best option modern medicine could offer. I came to the stark realisation that if I wanted to improve my situation, I'd have to take some ownership of my predicament and learn more about specific sports injuries. Hoping to find some books on sports injuries, I went to the Glasgow University campus, but soon discovered that you couldn't just walk into their library unless you were a student. Just down the street, however, the university bookshop was open to anyone, and I hid in a corner, reading sports medicine textbooks for a whole afternoon.

I was much further forward after this, learning that I'd likely torn a 'pulley' ligament that holds the finger flexor tendons tight against the bones of the fingers. Ligament tears like this tended to take a minimum of six weeks to heal, but more often several months. Healing would progress faster in response to gentle but progressive loading as the tissue slowly repaired. Although I could be out of climbing for months, I felt much better just knowing where I stood and seeing some light at the end of the tunnel. In the end, the injury took four months to completely heal, during which time I stayed away from climbing altogether so that I didn't get too keen and make it worse again. I returned to Dumbarton Rock expecting to feel very weak and unfit, but I was shocked and delighted to find that the physiotherapy exercises I'd done at home had made me even stronger. On my first day back, I completed the project I'd been stuck on before the injury.

Unfortunately, I suffered a string of these injuries over the next few years, which was a source of much psychological torment. It's hard to pinpoint the causes of them, and quite possibly there were multiple factors acting together. With hindsight, I suspect the most likely culprits were poor appreciation of good conditions for climbing, too much reliance on a crimp grip on small holds and poor diet. I spent so much time with injured fingers that I eventually abandoned my strategy of not climbing until they were almost completely healed. With a bit of experimentation, I found that I could still climb

Right: With Ali Weir at Dumbarton. Ali was a huge influence on my climbing, being up for exploring and ploughing through mileage on the less popular crags of the Scottish Central Belt. We climbed almost exclusively within an hour of Glasgow, but on a huge range of rock and route types. The loose, dirty and esoteric climbs we did together made me a much more capable and rounded climber.

with a very different way of taking the holds, using a grip called a three-finger drag, flexing only the last joint of the fingers, with the middle joint extended and locked straight. This relieves load from the pulley ligaments, and I could climb without pain. It was awkward at first, and I had to drop my grade, but it turned out to be one of the best things that ever happened in my training.

With a fresh and particularly bad pulley tear, I could tolerate only very easy climbing. Even warm-ups at the wall aggravated it too much. Another friend from school, Ali, was only just getting into leading on trad. He could climb about VS, had access to his dad's car and was great company. I decided that I would drop my grade all the way back to VS, which my finger could handle, but completely change the selection criteria for routes I'd try. Instead of going for the best routes on well-known crags, I'd specifically pick the most awkward, scary or loose ones on a much wider range of crags, including some very obscure venues hardly anyone ever went to. Ideally, this would test my confidence on intimidating terrain but not aggravate my finger. As it healed, I could gradually increase the difficulty.

The Scottish Central Belt, lying just south of the Highlands, has little crags dotted all over the place. Like the Highlands, they are massively varied in character. We climbed horrifically loose quarry routes on dolerite cracks. We climbed on sandstone so soft you could make your own holds. But mainly we just enjoyed climbing pitch after pitch at a different place every weekend. Sometimes I almost forgot that I was injured. The climbing, as planned, was mostly not on small holds but was often bold and on routes that needed cleaning as I went. The finger healed quickly.

After three months, I had gradually worked back up through the grades to E4 and felt far more solid, with a much better awareness of where my limits lay and of tactics to keep myself out of trouble while pushing myself on trad routes. I did my first E5 almost immediately after the finger recovered to full strength, and the following summer onsighted a long list of E5s all over Scotland. The process was transformative, and no amount of finger training in the climbing wall could have made up for the technique, confidence and tactical skills I gained. For the first time, I finally had what climbers call 'a leading head'. Strong fingers really matter for hard rock climbing, but they are just one ingredient. Without the other ingredients, they have no leverage in performance. Now, every kilo of force I could pull with my fingers went that much further on difficult routes.

Aside from Dumbarton Rock, another favourite venue after school was Craigmore, a ten-metre-high long edge with well-chalked-up climbs of all grades. It was about eight miles north of my house, on the road out to the Queen's View, and Peter and I could cycle there on the West Highland Way walking trail, which went right past the crag. At that time, if you were a climber in Scotland, you were a trad climber. The bouldering scene at Dumbarton Rock was a relatively niche part of climbing, and few climbers bouldered, although it was gaining popularity as a training method on indoor walls. Craigmore had a good scene, and on a weekday evening 20 or more cars would be lined up along the roadside.

The routes at Craigmore had only been given technical grades in my guidebook, as they were often top-roped rather than led, and some were short enough to feel more like bouldering. But leading was still a sought-after objective for the longest routes there. They were similar in nature to the gritstone edges of the Peak District in England, where short but very bold climbs were highly respected and in fashion at the time. The highest part of the crag contained the hardest route, Craig's Wall. The only British 6c there, it had had a few top rope ascents but had never been led. It was very convenient to try on a top rope, with a huge oak tree leaning over the top of the wall, perfect for securing a sling anchor. After attempts to work out the desperate sequence of tiny crimps, I would run laps on a much easier hybrid line, climbing up the easier first half of Craig's Wall and then stepping left into a crack line which circumnavigated the hard bit. When I was too tired to take any more of these endurance laps, I finished the sessions by jumping off the top of the route onto the rope. I'd stand on a small foot ledge and shuffle the anchor sling further out along the oak branch, such that the rope was well away from the wall and the fall was clean into space. Then I'd tell Claire or Peter to let out an armful of slack, and I'd jump off. It was totally safe, light-hearted fun, but I was acutely aware of the willpower required to commit to the jump, despite knowing that the rope would catch me after a few metres. As the sessions went by, I'd instruct whoever was belaying to let out two armfuls, then three, then four, until I was stopping only a metre or two above the ground on rope stretch. The plummets became much more exciting and gave us a bigger laugh, but I still wasn't quite over my reluctance to jump. The feeling of the ground being close at the end of the fall made it far harder to force myself to leap off the ledge, but familiarity with falling also greatly reduced the fear with every plummet. So I started asking Claire to add another half-armful, and then another 'wee bit', until I was jumping off the top of the crag and skimming the ground as I bounced on the rope stretch. It was a massive confidence booster.

At Dumbarton, I had already seen the value of consistency in multiplying the effects of being outside your comfort zone. It worked the same way here. No single exposure needed to be extreme. It could start small, and the increments could remain small throughout. But I'd seen that, for most climbers, the missing ingredient was simply maintaining the pressure and exposures for long enough to add up to seemingly remarkable progress. Now, I had no fear of dropping the length of a crag, so long as I had confidence in the system.

I'd been working systematically through the other routes at Craigmore, either leading or soloing them. After a handful of sessions, to my delight I successfully top-roped Craig's Wall. But after yet another heart-in-the-mouth jump off the top, I felt a tinge of sadness. Craig's Wall had never been led, so it wasn't an established, graded trad route. I had enough understanding of grades to know that this route would surely be an E7, which, of course, I'd never done. At that time, E7 was still the hardest grade

Right: Jumping off the top of Craig's Wall at Craigmore. Chris Wallace belaying with several metres of slack paid out, Claire looking on. © Ali Weir

in Scotland. I'd heard of every single climber in Scotland who had climbed at this level. Having successfully top-roped it, Craig's Wall would have been an obvious target for me to aim to lead. I thought perhaps one day it would be established as an E7, and then I might try to lead it myself.

But standing at the foot of the route, still tied on and having just leapt off, my perspective shifted. It clearly was an E7, and no doubt it would be formally graded one day by someone who had the confidence to lead it. What was stopping me from just doing it? The answer to this question was simply that new routes were things that other climbers did. Mainly the famous ones I'd heard of. This answer could not withstand further scrutiny. Rather like my hesitancy in jumping off the top of the crag on the safety of the top rope, I had no good reason not to lead Craig's Wall, other than irrational fear. It was no longer fear of not being good enough to actually climb the line, or of falling. The only fear remaining was the notion of new routing itself. As soon as I looked at the residual fear for what it was, it melted away and leading became the obvious thing to do next.

Opposite: Finally making it through the crux of Craig's Wall (E7 6c), on the first ascent.
© Ali Weir

At half height, Craig's Wall was split by a horizontal break which had a good flat hold and took a collection of tiny wires to protect a lead ascent. All were size 2 RPs, about 4mm thick, clipped to one rope. My first attempt made me realise that leading the climb was harder than top-roping it, and I fell, not even close to holding the crux lunge. Perhaps the anticipation of falling onto those tiny wires made me less efficient or more hesitant, but if this was the case, I wasn't consciously aware of it. I felt pretty confident after all the jumps off the top of the crag.

I heard a zipping sound, and a split second later, I was hanging from the last RP, with the other three spinning down the rope towards my harness. It was unnerving to be held by the single remaining runner, but it wasn't all that surprising as I knew that the last placement was the secure one of the four, and would never come out. Why did I fall? Likely the primary reason, which I was unaware of at the time, was the weather. As the difficulty of climbing progresses, friction on the holds becomes more and more important. Ideal conditions are a temperature as low as it can be without your fingertips going completely numb, low humidity and a breeze. But in mid-summer, with the trees surrounding the crag still in leaf, the friction probably wasn't as good as the day I'd successfully top-roped the climb.

Ignorant of the importance of climbing in optimum conditions, I returned for several sessions, having two or three lead attempts each evening and falling ten times in total. Each time, the first three runners would rip out and I'd be held by the last one. With all the airtime I'd had, I became totally unfazed taking the fall and was perplexed by why I kept falling. I was convinced that I was making small errors and would surely hit the holds right on the next try. On the 11th attempt, I wobbled my way to the top and clipped the sling on the giant oak tree. My first new route!

I was delighted with the climb itself, but in the months afterwards, I realised that the experience of doing a single new route changed my life and my whole approach to it. Before this, my climbing objectives were entirely limited to what was written in guidebooks. I had never really asked myself what was needed to put up a new route. The climbers who did so were often the very best, since lines on popular cliffs were unclimbed simply because they were hard. But I'd later learned that plenty of climbers who were far less physically able on rock than me were doing new routes all over Scotland, sometimes by the hundred, so it wasn't about ability.

There were other practical questions: How do you know if a route is unclimbed? Do you need to ask anyone before you claim a first ascent? I had already belayed Andy Gallagher on new sport routes at Dumbarton, and I once asked him if he consulted anyone before bolting them. He looked at me as if I was mad. 'I just did it,' he shrugged.

Right: With Claire after a successful ascent of my first new route, Craig's Wall.
© Ali Weir

At the time, I thought he must be very confident to have this attitude. But the more climbers I met, the more I observed that they seemed to share this approach. Rather, it was me who was the outlier. Gradually, I began to accept this way of thinking as normal. Knowing where to look for new lines seemed to come down to understanding the climbing scene and the crags themselves. By immersing myself in this and talking to other driven climbers, my perceived barriers to new routing dissolved.

New routing also filled me with a sense of abundance. I looked at my home crags in a different way, and unclimbed lines appeared in front of me. Now, whether a line had been climbed or not was interesting, but ultimately not that important. I didn't need to base my feelings about how to climb it, or whether I could climb it, on anyone else. I just made my own judgement. This habit spilled over into the rest of my life, too. I no longer viewed uncertainty as a source of discomfort. It later helped me carve out a life as a professional climber, supported by making films, publishing books, running courses and developing products. Most notably, later in life, it gave me the confidence to read and act on the scientific literature on health, training and nutrition directly, rather than through the filter of established institutions, which show considerable inertia in responding to new information. This has taken me to some unconventional places in my training practice but allowed me to solve health and performance problems that I'd previously thought to be intractable. It is hard to tell how much of this was just a process of growing up, and whether some other experience might have catalysed the same shift. Nonetheless, I felt far more confident to think and act independently and I cannot help but feel that climbing my first new route was the branch in the road where I started on this path.

All of this will no doubt seem obvious to some, and you might wonder why I draw attention to it. But for me, this shift in thinking was a big deal. Other people seem to start out with natural confidence to hold independent views, even if they fall outside the cultural norm, to live a different life from their peers or to lead other people with the strength of their own thoughts or decisions. A supportive upbringing where I was encouraged to push myself was somehow not enough. I needed to see it for myself, in an environment I could connect with more readily. I have often wondered if my dad's absence in my younger years delayed my progress in developing more confidence or decisiveness. I think I got there in the end, but I wonder if I may have stalled forever without the places and people climbing exposed me to.

Climbing was already a huge gift in my life, giving me enormous enjoyment and focus. But creating new routes unwrapped another layer that multiplied those rewards exponentially.

08.

HEAVY TRAINING

Left: Flashing Shine On (E7 6c) at Stanage Edge, Peak District in 2000, with Richard McGhee belaying.

Around the time I climbed Craig's Wall, I was 19 and had not long left school. Although I had attended much more in my final year and left with six Highers, they were spread across two years, which meant I didn't have the grades I needed to get into university. I was unsure if I wanted to go down this route anyway, although the Education Officer's comment that he thought I'd fit in well at university possibly did weaken my bias against formal education. I might otherwise have dismissed the idea altogether after my experience of school, which sometimes struck me as being geared primarily towards instilling compliance and helping you find your place in a hierarchy, with actually learning about the world taking second place.

A friend worked as a postman and told me there was a job going at his sorting office, so I took it and quite enjoyed the meditative simplicity and solitude of walking the streets in the early morning delivering mail. But my lack of discipline for going to bed early eventually took its toll, and I switched to another job stacking shelves in a large branch of Boots on Sauchiehall Street in Glasgow.

I struggled in this job for a reason that I felt awkward about and never really shared with anyone because I thought I was just weird. We had rigid 15-minute breaks in the middle of our shifts, and between these I'd get ravenously hungry and feel shaky and weak, almost like I could faint. Throughout my late teens, I'd been steadily gaining fat around my waist, which I was intensely embarrassed about. Claire would sometimes say that she found it cute that my T-shirt would hang off the end of what was quite a noticeable 'pot belly'. This made me want the ground to swallow me up. I desperately wanted not to be fat. Hard rock climbing was very dependent on strength-to-weight ratio, and, almost without exception, the best climbers were as lean as stage-ready bodybuilders, although, unlike bodybuilders, they were also wiry with skinny legs and only modest amounts of upper body muscle.

Like many young Glaswegian teenagers, my diet was pretty poor, but I didn't really grasp why this would matter. As far as I knew, body fat was completely under voluntary control, simply by manipulating energy balance. It was just a matter of how many calories you ate and how many you burned. If you had gained excess fat, it was because you'd let yourself eat over and above your energy expenditure, and to lose it again, you'd just have to go a little bit hungry for a brief period. The problem was that I was hungry every waking moment. Sometimes I was so unbearably hungry that I couldn't concentrate on anything except how long it was until my next meal. It hijacked my attention constantly and really got in the way of enjoying life. I'd witnessed other climbers who also had this problem, but only because they were forcing themselves to be underweight in the hope that they could climb better. I felt hungry even when I was significantly overfat and gaining weight. Confusingly, it didn't correlate all that closely with how much I ate. I thought I must just be greedy, and I hated it.

For breakfast, I'd eat Weetabix, a cereal that comes in blocks. Most people would eat two blocks. I'd eat five and have to stop myself from adding more. All I had to do was get the train for 30 minutes and work for two hours before I'd get a 15-minute break and could go to the canteen and eat sandwiches, but an hour into the shift, I'd be ravenously hungry and feel awful. It felt like a fight to make it to the break. I started buying a chocolate bar on the way to work and would eat it while sitting on the toilet just to keep myself going. I wasn't proud of this, but the hunger was real.

Gradually, I improved my diet and removed a lot of the junk food from it, which did bring my weight under slightly better control. At the time, I judged that this was because of the calories in the food. I now suspect it was more related to the type of food I had been eating, specifically the combination of starch, sugar and vegetable oils that characterise most Western junk food. Overconsumption of calories was a symptom rather than a cause. Even this was only enough to prevent further weight gain, so I also went on some quite severe diets. When I was 16, I had been desperate to climb a sustained traverse on the Dumbarton boulders called 'Consolidated', which had only a couple of repeats from the best local climbers. I'd been trying it on and off for weeks but couldn't get past the hardest section, about five moves from the end. It was obvious to me that the excess fat I'd gained was not helping, so I decided to fast for a few days. For four days I ate just an apple and a cup of tea each day. Curiously, I noticed that, although I was hungry, I didn't have that wobbly, shaky feeling or the sense of absolute urgency to eat. For large parts of the fast days, I actually felt less hungry than while eating normally. This was a fundamental observation, but since I didn't have any framework to understand it, it remained an unresolved curiosity.

Above: Pulling on quickdraws on Natural Born Drillers (6c+) at Dumbarton Rock. During this period, my weight and my climbing standard fluctuated considerably. Cameron Phair and Andy Gallagher had just bolted and climbed the route and invited me to try it. I failed to onsight it, taking a big fall from the last move, and had to return and redpoint it. With a diet change, my next route was a repeat of another of Andy's new routes, Dum Dum Boys (8a+) - a jump of eight grades.

On the fifth day, I got back on Consolidated and made it to the last move. Immediately I did a second round of the four-day fast, and the following day, I completed the route. After this, I rapidly regained all the excess fat, but I just thought I'd been greedy again and continued to manage my weight with less severe periods of calorie restriction. It was a constant source of frustration and even shame that simmered in the background, since I had no better solution. I wouldn't understand what was going on for many years to come.

After my shifts at Boots, I'd buy lunch and get the train straight to Dumbarton Rock and climb for the evenings. Gradually, I started running out of established climbs to point myself at, so I began working on some unclimbed projects. I also finally worked up the curiosity to try the moves on Requiem. Dominating the skyline above the boulders, literally leaning over the top of them, it was impossible to ignore. The smooth orange wall lit up in the evening sun was a constant reminder of my original climbing goal. With Requiem in my face on an almost daily basis, I had to routinely decide I still wasn't ready to try it, which became tiring. I had now done as many climbs on the boulders as the best local climbers, and I could no longer hide from it, so I put a rope on it to try the moves. It was just easier to get on with it than to stare at Requiem from the boulders every day, for another year.

Like the climbers I'd seen on my first day climbing, I too hung on the rope, contorting my fingers into the crack and struggling to make a single move. The terrain was very different from the boulders, and something about the face felt intimidating, even though it was only 120 feet high. As I had done on the boulders, I whittled away and worked out the moves over several sessions, before coming up against a brick wall. I could link the moves in three sections, but repeatedly failed to improve on this. Obviously, the climb required not only strength and climbing skill but a lot of endurance as well, something I naively felt I already had. Much later, I would come to grasp that the other endurance climbs I was doing at that time were not, in fact, true endurance climbs. They had short, hard crux moves that I could manage because of my bouldering strength, and the rest would be easy enough that I could scrape through. Requiem was a true endurance climb, requiring the ability to recover on better holds but still on steep terrain. It was essential to remain fresh all the way to the crux, which was the very last move, slapping for the top of the crag in an incredibly exposed position.

Frustrated, I had a sense that I was missing something. Possibly if I'd been climbing with a wider net of climbers at a higher level than me, I'd have figured out where I was going wrong a lot sooner. But I still mostly climbed on my own or with a small group of friends from school who were more casually involved with climbing. I had read every

Top right: Resting halfway along the traverse of Consolidated (Font 7B+) at Dumbarton Rock in 1996.

Bottom right: At Dumbarton Rock. © Claire MacLeod

Top right: Attempting Steall Appeal (8b) at Steall Hut Crag on my first visit to Glen Nevis, a place which immediately kickstarted a love of rock climbing in the Highlands.
© Claire MacLeod

Bottom right: On Cut Throat (VI,6) on Beinn Udlaidh.
© Cubby Images

Above: At the top of Dalriada (E6 6b) on The Cobbler, after the second ascent in September 2000. The previous evening had been my engagement party with Claire and all my friends and family. I put down my whisky at 7am and caught the train to The Cobbler with Iain Hutchinson, convinced that it would be the last day to do it that season. I was inspired by a brilliant picture of Gary Latter on the first ascent but surprised to find that the route had nine pegs on it. I intended to climb the route without the pegs, which would make it E8. But with sleep deprivation and a severe hangover kicking in, I ended up clipping one near the top, something I later regretted. The combination of strength from bouldering and rehearsing moves on a top rope first facilitated a 'clean' approach to mountain trad, without fixed or pre-placed gear.

piece of information there was to read on training for climbing at the time, but I lacked the experience to properly assess my own weaknesses.

Claire and I had been talking about the idea of studying again. We'd seen an 'access course' run in the evenings by Glasgow University that was like a primer for people who couldn't get to university via the normal channels. Claire suggested I could go on to study sports science since I was obviously grappling with how to train properly for climbing and seemed to have all sorts of problems to solve - training protocols, weight, injuries and tactics. It made sense, and I enrolled. Within a month, thanks to the tutor who taught the chemistry element of the course, my entire attitude to study and academic learning had turned on its head. There was something about the manner in which my tutor spoke about science and how it related to the real world that just seemed to click. He simply enthused about the wonder of chemistry and how it makes the natural world fit together and dictates its characteristics. I came to really enjoy my studies and I easily passed the exams needed to start a degree in physiology and sports science the following year.

As my school Education Officer had predicted, I loved university life. The learning was interesting, the students were fun to be with and the flexibility of the routine fitted in perfectly with climbing. The fact that I'd had a couple of years out of school and worked in different jobs seemed to have helped me as well. It gave me an appreciation that the work required of us as new students wasn't difficult if you just attacked it, and the alternative to doing so was returning to those same jobs. I actually liked both jobs I'd had and would do them again since they were light-hearted and relaxing in some ways. It was just the inflexibility of them that I couldn't handle. If the sun was shining, I had to climb.

Through the winters, the sun in Glasgow didn't shine, however, and I spent a lot of time training in Glasgow Climbing Centre, a big wall in an old church. I'd never been good at climbing indoors. My fingers sweat a lot and I have always felt I cannot apply my full strength on artificial walls, so I struggled to keep up with other climbers who were going well. This was great for my training, though, and what I lacked in friction on the holds, I made up for by trying really hard at all times. I never rested on my laurels when training indoors, because the minute I did, I'd perform disastrously compared to other people I climbed with.

The most important factor for any athlete to progress is the ability to try harder than almost anyone else and to be able to do that consistently. This seems so obvious, almost a platitude. Despite this, real athletic improvement remains elusive most of the time; it seems fleeting, sporadic and difficult to attribute to a specific change in

Right: On the first ascent of In Bloom (Font 7C+) at Dumbarton Rock in 1998, traversing the rail rightwards into the crack of Pongo. The sit start to Pongo was another unclimbed project I was trying at the time, climbed later the same year by Malcolm Smith.

training. It ought to scale with the ability to grind; to complete more sets and pull harder than the next person. Yet it doesn't. To my mind, this is because 'try harder' means different things to different people. There is a place for grind. It is absolutely necessary at times, but almost never sufficient. Those athletes who put their motivation purely into grind neglect the rest of the picture. The athlete also needs to try hard to recover, to try hard not to apply too much of the same type of stress per unit time. They need to try hard to understand their technical mistakes and try hard to find real inspiration and thus sharpen the kick from every training and recovery session.

This widening of the dimensions in which you can try hard, as well as the act of trying hard itself, are all trainable skills. Any environment that nurtures this is like gold dust. During this period in my climbing, both my physical and psychological environment seemed to be working in sync to help me along at a steady rate. The physical effort it took me just to barely keep up with the best climbers at my local wall must have been a strong training stimulus. I later found out that it gave me huge leverage out on the real rock, where cool winds and friction meant I could translate every ounce of strength I had into pulling power on the holds. In parallel, my doubts about my ability to handle arduous physical training helped me cultivate a keen sense of my state of readiness to train and to err on the side of better quality rest. This combination of generous effort during sessions and generous rest was highly effective.

I'd keep climbing until the wall closed at 10pm. Each night, at 9:55pm, they'd play 'Bohemian Like You' by the Dandy Warhols on the speaker system, and for the last few seconds, they'd turn it up really loud and then turn it off. The whole routine was meant to be a polite warning to finish your last climb and pack up so as not to hold up the staff. But this was lost on me. For years I only noticed it was time to stop climbing when they turned the lights off. But every time I heard that song on the radio or playing in a shop, I'd suddenly feel an inexplicable sense of anxiety. One day, long after I stopped climbing there, I heard the song on the radio and felt the same sense of stress, but this time I realised why, as my brain finally connected it to 9:55pm in Glasgow Climbing Centre. I felt pretty stupid.

The following spring was the year 2000, and the general mood of society seemed to be one of optimism. This matched my own vibe when I emerged from the climbing centre in spring and returned to Dumbarton. The hardest climb on the boulders was the sit start to a thin overhanging crack called Pongo. I had tried it a little before, while it was still a project. Malcolm Smith had since made the first ascent and given it a grade of Font 8A - an almost mythical grade at the time, at least in Scotland, where only a couple of climbers had managed to climb at this level. Straight away, I could do the crux move and I quickly repeated it. I had taken a step up. Time to get back onto Requiem.

The year before, I'd sensed that I was not far from a 'threshold' of strength that would allow me to link much longer sections of Requiem. Scrambling to the top of Dumbarton Castle to hang my top rope already felt like doing something special, and the anticipation of discovering if my winter training had paid off added an extra layer of excitement. On my first try on the top rope, I cruised past my previous high point, and after a bit more familiarisation, linked the whole thing on the top rope. I had reached a level of basic strength such that many of the moves didn't feel hard any more, and on top of this, I'd built some stamina over the winter. Still, leading the route would be intimidating, even though it is mostly well protected.

With Ali belaying on the ledge below the crack, I set off and wobbled my way up it, over-gripping all the way, falling on the crux final move. When the rope caught my fall, I threw my arms in the air and hung there, screaming with excitement. It didn't matter at all that I hadn't done it. It would still be here tomorrow. The point was, I'd realised that my dream since day one of climbing was imminently possible.

A few days later, on a breezy evening with many others at the crag, I set off on Requiem and sailed to the top on a wave of pure joy that I had gained the ability to actually pull off such an ascent. In the cool wind rushing up the Clyde estuary, the

Top right: On the second ascent of Pongo Sit Start (Font 8A) at Dumbarton Rock in 2000. I'd failed to climb it for a few years, but after shedding some of my Western diet weight, I could run laps on it.
© Nick Tarmey

Bottom right: At the top of the crack on Requiem (E8 6b) at Dumbarton Rock. At this point, the crack fades out and the route breaks right to follow a flake line to the top.
© Nick Tarmey

Opposite: The upper section of Requiem.
© Nick Tarmey

holds felt like Velcro and my arms were not tired at all. The overwhelming intimidation of just climbing on lead on that wall was also completely gone, and I felt utterly at home. As I pulled over the top, I could hear climbers below clapping and whooping in celebration. An ascent of Requiem was still a very rare sight at that time, and many of them knew what it meant to me personally to climb it.

I have often wondered how things would have worked out if I hadn't seen those two climbers trying Requiem on the first day I went to Dumbarton Rock, seven years earlier. What if they had been on one of the much easier routes? As a novice, the sight of any rock climber was incredible to me, and perhaps I'd have stood in awe and ended up with a much easier goal instead. I can't say for sure. But the fact that I had naively told myself that I wanted to climb that route, whatever it was, had been a driving force in the background ever since. Having identified with it, I couldn't later erase the thought of it. There was no going back. Its effect became a constant voice in my head: *If you don't force yourself through this move or this climb today, you'll never get to Requiem.* It was a gentle, insistent push, with everyday small excursions from my comfort zone cumulatively having a huge effect on me during the seven years I'd been climbing.

Sitting on top of Requiem in the roaring wind, I stared at my hands and forearms and absorbed the reality of having done the climb and that it had felt so effortless in the end. How did I get here? I immediately wondered if I could continue with the same approach of identifying with other seemingly unrealistic goals. Just because I couldn't imagine being good enough to do them right now meant nothing. Seeing that they were possible only came at the very last moment.

Soon afterwards, I was sitting outside university on a sunny evening with friends from the mountaineering club, discussing Requiem and other climbs I might try next. As we talked, I noticed a feeling I'd never knowingly experienced before. It only lasted a brief moment, although its memory left an imprint in my mind. It felt good, but at the time I wouldn't have been able to describe it. Today, I would call it confidence.

THE IRON ROAD

09.

Previous: The first ascent of Flook Talk (E8 6c) at Dalbeg, Isle of Lewis. An initial onsight attempt on this route ended in a gear-ripping fall below the crux. A fall from the crux requires the belayer to sprint backwards to take in enough rope to keep the leader off the rock platform below. I returned with Claire and didn't fall off.
© Cubby Images

During my early years at Dumbarton Rock, the best climber was a Glaswegian called Andy Gallagher, whose wonderfully intuitive and dynamic style reminded me of Johnny Dawes, the world-famous English climber who had opened Britain's first E9, The Indian Face, on Clogwyn Du'r Arddu in North Wales. Andy's creativity also extended to envisioning new routes. He embraced sport climbing and, despite this still being controversial in Scotland in the early 1990s, with its deeply embedded trad climbing ethic, he bolted many new routes at Dumbarton and the surrounding crags. One morning I turned up to see Andy drilling bolts on the smooth slab directly below Requiem. I hardly knew him and was delighted when he appeared around the side of a boulder and asked if I would belay him on it. With the bolts just placed, he jumped onto the route and danced up it, creating a 7a+ called Persistence of Vision. His climbing style was so different from mine; he pieced the sequence of movements together in a much more fluid manner, making quick adjustments on the hop and using methods I wouldn't have thought of.

Persistence of Vision finished at the start of the overhanging wall, just left of Requiem, but Andy said it could continue into a much more futuristic line up the face beyond. It seemed unimportant to him whether it would be actually possible, as he enthused about the aesthetics of the line itself and the excitement of trying it. Later, Mark McGowan, another great climber of that period known to everyone as 'Face', tried this second pitch up the overhanging wall. Mark was a very bold climber who had done some incredible free solos. He ran an outdoor shop in Glasgow, and whenever I'd go in there and chat with him, he had a very calm and level-headed demeanour that made it easy to imagine his psychological control while soloing.

Although I didn't know either of them well and would never have dared ask if I could climb with them, I always observed their respective approaches carefully and tried to notice details of the tactics they used, either practical or psychological. On the line left of Requiem, Face could make barely any moves at all. Yet when he lowered off, he seemed unmoved and described the prospect of the route to the audience below in a very matter-of-fact way. The line remains unclimbed to this day. It struck me that it took a degree of confidence to attempt something futuristic and unclimbed, but perhaps even more confidence to take failure in your stride and not feel a sense of impostor syndrome. These guys appeared to accept that if no one actually tried these routes, no one would ever know if they were possible. It's very simple to say this, but somehow it seemed like a different story to just give something a go, instead of leaving it to someone else.

Dumbarton Rock is a pretty unusual cliff in Scotland in that traditionally protected climbs and bolt-protected sport climbs coexist right next to each other. The relationship isn't perfect, and sometimes one style encroaches on the territory of another, but by and large, it seems to work. The rules of what should remain trad and what should get bolted are fluid and have changed in the time I have been a climber. In general, though, there are no hybrids: a route is either all trad or all bolts. The lines either side of Requiem were both obvious targets to bolt, but I wondered if it was possible to climb two other sections of the face which were not completely independent lines. On the left, a striking crack graded E5 curved leftwards, eventually becoming almost horizontal. It was called 'Chemin de Fer', the iron road, likely owing to the ancient pegs and wooden wedges still lining the crack from the days of aid climbing.

If the wall above Chemin De Fer could ever be climbed, it would need to be done without bolts, since it would be accessed via the crack below. This would actually make it a really exciting trad route, since you'd have good protection up to about 70 feet, followed by a runout of 40 feet or so on the headwall above. Perhaps only the last couple of moves below the top would be truly dangerous. So the issue would very much be whether the moves were feasible. So far, no one had managed to force a route on the wall away from the crack lines, whether using bolts for protection or not.

Influenced by Face's and Andy's relaxed attitudes to looking at these lines, I decided to abseil down the headwall for a look. Aesthetically, the most appealing line was to leave the Chemin crack and head up and rightwards across the wall towards another line of better holds much higher up. Practically, though, it didn't work. Hanging on the top rope, I played on the moves for several sessions. I could see it was possible to go in this direction initially, but after 20 feet, the holds pulled you back left. Eventually I decided this would have to be the way the route would go. There were two moves I couldn't do, and even the others seemed right at my limit, with much harder climbing than Requiem.

On a warm June day, I finally grasped the importance of good conditions for attempting climbs truly at your limit. The second crux move involves crossing under to take a smooth, flat sidepull in front of your chest. The hold is completely useless when your fingers first reach it. But as you place your fingers on it, you must pull in hard with the left foot and spin your body round from left to right. As you spin across, the sidepull becomes usable, and if your timing is correct, you stay on. On a previous session I'd worked out the move and had done it in isolation a few times, but this time I couldn't pull on the hold at all. Staring at it in frustration, I saw four dark fingerprints of sweat against the white chalk. Soon afterwards, tired, I told my belayer that I'd be coming down after one more try. Just then, the sea breeze picked up and the air, which

Left: Falling off Tunes of Glory (E5 6b) at Traigh na Beirigh, Isle of Lewis.
© Cubby Images

had been stagnant, suddenly felt fresher. Despite the tiredness, I pulled on again and climbed through the move first try.

After this, I felt rather more relaxed about failing on moves on hard routes. On countless occasions since, I've learned that in different conditions, it can feel like you've been given twice the power in your fingers. The truth, of course, is that you've just been given the friction through which to apply the strength you already have.

By the end of June, I had managed to link the whole line on the top rope. I had to stop and take stock. Over the past year, I'd been climbing hard trad routes in other areas of the UK and had repeated several E8s. It was obvious that there was no trad route in Scotland, and very few anywhere, with anything like this level of physical difficulty. It seemed clear to me that this project could never be an E8.

Trad routes vary immensely in character, and it can sometimes be fiendishly difficult to estimate a grade. The overall grade is first and foremost an amalgamation of the physical difficulty of the climb and how dangerous it is, together with a host of lesser factors like length and logistics. Really poorly protected and dangerous climbs can be the hardest to grade, as they are dependent upon how much trust a climber places on very subjective assumptions. For example, if the protection is a poor nut in a flared crack, one climber might judge it has a high chance of catching a fall, while another might view this protection as 'psychological only'. Or if a ground fall is the likely outcome, a young climber without a fully developed prefrontal cortex might judge that a 25-foot fall onto a grass slope with a few rocks sticking out is nothing much to worry about, whereas someone else might see a career-ending accident. Climbs like the project at Dumbarton were more predictable. The gear was beyond question. There was no way you'd hit the ground, but it would be a long fall and definitely exciting. Routes of this character with a sport climbing grade of 8a or even 7c+ were given E8. Therefore, at a likely 8b physical difficulty, this project must be E9.

Since climbing Requiem and trying to lift my sights a bit higher, I had set E9 trad and 8c sport as my dream lifetime goals. Neither grade had yet been climbed in Scotland, and given how strong and bold the best Scottish climbers were at the time, these seemed like a ridiculous stretch. Although I'd already climbed the project once on a top rope, I was nowhere close to being able to lead it, which always feels that bit harder on trad climbs. So I had a rough plan to keep pushing my general trad standard until something changed.

Shortly afterwards, I made my first trip to climb on the isles of Lewis and Harris in the Outer Hebrides with two good friends, Niall McNair and Steve Richardson. We'd been invited along by Dave Cuthbertson and his wife, Joanna George. I had been delighted to finally meet Cubby and Jo a couple of years earlier on a trip to Glen Nevis, and we had kept in touch. Both seemed like gurus of Scottish climbing, and when I visited their house in Ballachulish, I quizzed them endlessly. They were both great mentors and offered a lot of encouragement and advice. Well before I'd climbed Requiem or anything like it, Cubby was always straightforward in his encouragement, and I took it as a valuable opportunity to listen to what he had to say about my own climbing and what I could do to improve it. He noticed that my motivation and willingness to repeatedly try routes were assets that I could leverage. He readily pointed out that I was carrying a lot of excess fat and said, 'If you lose a couple of stone, you'll be flying.' I was reassured that this matched my own perception. He also suggested that I'd benefit from getting away from just climbing at Dumbarton and gathering more varied experience. I thought I'd already done that during my injured period, and this shifted my perception a bit. Perhaps this was why Cubby and Jo invited us to join them on Lewis.

Possibly the greatest thing about Scottish rock climbing is the variety that exists across the country. I've never yet visited anywhere else with such diversity of landscape, rock and character within such a compact geographical area. As soon as we rolled off the evening ferry onto Harris and drove up across Lewis, I was really taken with the place. It felt like another planet. Contrasting deep brown peat bogs and wind-scraped slabs of silvery gneiss stretched into the distance over rough, rolling hills. The colours of almost everything - sea, beaches, peat, machair and rocks - seemed extra vibrant. Despite the powerful first impression of raw natural beauty, signs of a long cultural history were everywhere, from the Gaelic language still spoken in the pubs and shops to the ancient crofts, lazy beds and peat stacks that line the lonely roads stretching across the bogs.

We spent the first days exploring the sea cliffs of Lewis, and Cubby showed us several excellent cliffs he'd been gradually developing over recent years. Steve, Niall and I were also keen to see a cliff we'd heard a lot about: Sron Ulladale, in the mountains of Harris. In Gaelic, sron means nose, and the cliff rests on the end of a long hill ridge, which the last glacier cleaved one end off to leave a 600-foot leaning face that overhangs its base by 150 feet. The scale and steepness of this cliff make it one of the most striking features in Scotland. The fact that the Sron is also hidden in the middle of the hill range and not visible from the road adds to its mystery. Many of the best trad climbers in Britain had travelled here to attempt new routes, including Doug Scott, Johnny Dawes, Paul Pritchard and Cubby himself.

Jo and Cubby dropped us off at the road end near Sron Ulladale on a good forecast

Right: First ascent of Cnippy Sweetie (E7 6b) at Traigh na Beirigh, Isle of Lewis in 2001, with Niall McNair belaying.

and agreed to pick us up again after a couple of days. It had taken most of the morning to get hold of a forecast and then drive down from Lewis, so it was a bit late to start a route, but at least we could walk in to have a look. As we came over a pass, the massive overhang in profile was just as impressive as its reputation suggested, and I could see why it had attracted the very best climbers for generations. Beyond lay a seemingly endless expanse of rough moorland resembling a barren lunar landscape, apart from the fact that it was dotted with countless lochains and bogs. It would be an intense place to climb.

With the huge Sron towering over us, we could hardly stand still with excitement, and the question of whether it was a good idea to start a route was brought up again. The correct answer was almost certainly 'no', but with talk of long daylight hours in June this far north, 'no' somehow became 'yes', and we got the ropes out below a six-pitch E5 called Stone. It was 6pm. Although all three of us were competent trad climbers, Niall and I both having onsighted many E5s and Steve soon to do the same, I had never climbed a proper multi-pitch rock route before. Given our collective ability and my winter climbing experience, the route choice wasn't outrageous, but I suppressed a niggling thought that it was not the most sensible either.

We made steady progress and reached the base of the crux fourth pitch some time after 9pm. The pitch above looked spectacular, very steep but well-featured with cracks, the sort of climbing that makes hard trad onsighting so much fun. The trad protection and steepness can be intimidating, but the climbing can feel okay if you are fit and strong. It was Niall's lead, which I was envious of, not least because of the freezing wind cutting across the exposed belay.

Niall is extremely flexible, and I watched him make some dramatic shapes bridging up a grossly overhanging groove and then try to arrange gear in a bulge above. He wrestled through the tricky-looking bulge and steadily advanced up the exposed skyline, reappearing as a small dot far above. The description had said to belay on a large ledge above the groove, but Niall seemed to be having difficulty and was perched awkwardly, well left of the ledge. Even more oddly, we could hear his voice.

Niall wears two hearing aids and partially relies on lip reading. Therefore, once he was more than halfway up a pitch, he could shout down to us, but there was no point trying to converse. So he was usually silent apart from either grunts or whoops of excitement. These one-way communications were quite binary, and I appreciated them since it was easy as a belayer to monitor how he was getting on with the climbing. Generally, he would alternate between shouting 'Watch me' if things were challenging, and 'Who's the daddy?!' if things were going well. His current chatter unsettled me.

It was getting late, and a strengthening wind brought a sense of anxiety that we needed to be closer to the top. Tied to one of the two lead ropes, it was Steve's turn to climb next. As he moved upwards, I realised that all the runners were clipped to Steve's rope in the first two-thirds of the pitch. He'd taken them out as he went, and my rope hung nearly 100 feet without being clipped to anything. If I fell seconding, I'd swing miles out into space. I had no equipment with me for jumaring, and my inexperience at this sort of climb suddenly seemed to matter quite a lot. I was also, by now, shivering badly.

When Steve joined Niall at the belay, he too seemed to take an awfully long time to get established at the anchor, and there was some animated discussion between them that worried me. Finally, a shout came for me to climb. Moving up the groove, I felt like a cardboard cut-out, I was so cold. My hands were completely numb, and the chalk seemed to slide off my fingertips as if they were made of polished leather. At the bulge, the wind raged upwards. I had partially unzipped the small duvet jacket I was wearing, and it blew up right over my head and in front of my face. All of a sudden, numb hands, glassy sloping holds on the crux and temporary blindness seemed to conspire against me, and I felt very close to falling off.

As I looked down and stabbed at little smears with my toes, I watched in horror as my rope started to drop in a huge loop of slack. Niall had clearly let go of his belay device, and the skinny rope was just running through it. Peering over my shoulder at the rapidly growing loop that was now catching the gusts and blowing around, I realised that a fall would be absolutely wild. Goodness knows how far I'd go before that loop came tight. Even once it did, I'd be going so fast, Niall wouldn't be able to get a hand back on the belay plate, and I could drop the whole rope length onto the belay anchor. I felt sick. *I'm meant to be seconding!* Slapping my way around the bulge, I screamed, 'The fucking rope! Take in the fucking rope!' But it kept dropping, the base of the loop now at least 30 feet below me. What the hell was going on? I had no option but to climb on, and as I approached the ledge, the rope finally started bobbing upwards rapidly. I heard Steve shout, 'Wait till you see what's up here!'

The ledge was a scene of farce. Niall and Steve were leaning off the far left edge, practically sitting on each other. At the other end sat a large, fluffy golden eagle chick, looking distinctly unimpressed. Golden eagles have nested on the Sron or in its vicinity for a long time, and we had checked where they were nesting beforehand. We'd been told that the nesting site had been on the other side of the mountain for the past several years - obviously out-of-date information. We needed to leave as fast as possible. Without saying anything, Steve loaded my harness with the rack as I

Top right: Niall McNair approaching the crux bulge of Stone, the ledge with the eagle's nest on the skyline above.

Bottom right: Niall on Lewis.
© Cubby Images

Right: Starting up Stone (E5 6a) on Sron Ulladale. © Steve Richardson

clambered past him, still shaking from the previous pitch, and I continued upwards, at least now warmed up.

There is nothing quite like the sight of an adult golden eagle flying right past you on a cliff. We aren't used to seeing such large and impressive birds in flight at close proximity. While Niall and Steve were on the belay, the mother eagle had flown past a couple of times, and Niall had feared it might attack him. In a bit of a panic, he had indeed dropped the rope.

As I sped up the final pitch, the setting sun emerged for a few minutes and illuminated the upper 300 feet of the cliff in deep orange and purple rays. Meanwhile, the base of the glen was cast into deep shade and appeared like a bottomless black void, adding to the other-worldly atmosphere. I placed a hurried runner and glanced round for a moment to admire the immense display of light. At a ruffling sound beside me, I turned further to see the eagle float gently past, lit up by the sun. I wanted to stay and watch as it soared in the updraught, but I allowed myself only a couple of seconds before racing up the easy final metres. This moment cemented my relationship with the Hebrides, a place I have visited every year since. The next day we returned and looked from across the other side of the glen to see the eagle flying back and forth to the ledge and her chick.

I climbed a lot with Niall and Steve that summer, onsighting many E5s and E6s. Although I was stronger on the hardest moves, they had more natural and efficient technique than me, and Niall in particular had a lot of endurance. I didn't fully appreciate at the time how big a weakness endurance climbing was for me, and I struggled hard to keep up with those guys. The volume of time I spent both pumped and scared a long way above protection was excellent training for me to try an E9.

Steve is a very solid climber, composed under pressure, even when things get really out of hand. Although extremely determined, he anticipates his limits very carefully and so tends to choose routes at the right moment when he is ready. However, when doing back-to-back E5s in the mountains, some of which hadn't been climbed in years and weren't very clean, things didn't always go according to plan. On the way home from Lewis, Steve, Niall and I stopped off at Creag a' Bhancair in Glen Coe. The weather was pretty poor, but because of its size, the wall tends to remain dry if the rain is only light. Between showers, Steve started up a very serious and rarely climbed E5 5c, which has one runner low down, then a massive runout for the remainder of the 20-metre pitch, such that the last 12 metres or so are just free soloing trailing a rope. The route gets soaked by wind-blown drips in the winter months, and some mossy holds made progress slow. Steve pressed on but looked uncomfortable. When close

Right: Niall taking a fall at the crux of Cosmopolitan (E5 6b) in Glen Nevis. © Cubby Images

Right: Discussing moves with Niall in Glen Nevis.
© Cubby Images

to his limit, he rarely says anything but just tends to go red-faced with effort, and his normally calm expression tenses into a grimace. At the worst possible moment, when he was completely committed near the top, another rain shower came on, heavy this time.

There was nothing either of us could do except watch in horror as Steve pushed onwards. The holds were getting wet, and rainwater poured off the back of his helmet. As the belayer, I was gearing up to sprint down the hill as best I could if he fell, with virtually no chance I could have stopped him from cratering. With each slap upwards, he quietly repeated the words, 'I'm dyin', I'm dyin'!' It was a fair assessment of his battle for survival, but he made it to the belay, and that was enough fear for the day.

In contrast, climbing with Niall was like this almost constantly. Niall is quite an extroverted person and liked to build up to an almost fearless approach by just onsighting lots of routes and talking up his confidence. Although effective much of the time, I worried it could all go badly wrong at any moment, perhaps because I have such a different personality. Positively reinforcing my own ability like this always seemed counterproductive. I couldn't really see how it could be relied upon to keep me safe. To me, runners and holds were what kept you safe and telling yourself, 'I'm ready for this' felt pretty pointless, even delusional.

However, maybe it was me who had the wrong approach. My first trip with Niall had been a couple of years previously, to the Peak District in England. Desperate to get on routes as hard as possible, we spent a week at The Roaches, where Niall and I both onsighted a couple of short E6s. I top-roped an E8 slab called Obsession Fatale and could climb it first try. This doesn't mean much, as I soon discovered. The route has no protection and, encouraged by my consistency on the top rope, I considered going for the solo, which would have been my first E8. It seemed easy for such a huge grade, although the last move was precarious. We had seen a group of lads earlier in the day, and Niall went off to look for them to ask if we could borrow their bouldering mats, during which time I nervously paced about at the foot of the route, asking myself, *Am I really ready to solo E8?* Niall came back with the lads, and they eagerly dropped their mats then ran round to the top of the climb and leant over it, exclaiming how far the landing was below. 'Wow, you'd be a total sandwich if you came off that top move!'

Finally, I chalked up and started climbing. With every move, I felt worse and worse. Nothing was objectively wrong, yet inside my head, I was falling apart. I was desperate not to waste the opportunity to attempt the route with the mats for protection, and I continued up the first couple of delicate moves, arriving at a good foothold before the crux. Being an easy-angled slab, I could afford to hesitate, and I did. Something screamed at me not to move any higher. Another older climber who seemed to know a lot about the hard routes in the area had stopped by to watch before I'd set off. After seeing me hesitate and then start to wobble, he eventually intervened and gave me a way out. In the tense silence, I heard him say, 'Are you sure this is a good idea?' Of course it wasn't. I reversed the moves and told the spectators that I didn't feel good on the route.

Intervening like that when a leader is in a serious position is not an easy thing to do or to judge. A bold lead takes complete mental focus and incredibly strong resolve to push on. The time for processing doubts is in the preparation, on the ground or on the top rope. Of course, things can go wrong on a lead or solo, and quick thinking and decision-making are still needed. But this is a last resort and definitely not something to rely on regularly. A rehearsed lead should really be about execution, with doubts never allowed to dominate. I didn't yet have enough experience to recognise that I was not fully prepared, and didn't anticipate how different it would feel to trust the delicate smears without the safety of the rope. I'd left myself open to absorbing that new sensation at the worst possible time, when the danger was real. However, no preparation is ever perfect and just because a leader is hesitating doesn't always mean that retreat is the best option. For one thing, they'll likely be back at the same

Left: Attempting Velvet Silence (E6 6c) at Black Rocks in the Peak District, with Michael Connor spotting. I had just bought a small crash pad and was the first of my peers to own one. Michael brought a puffy futon mattress in a zipped storage bag. When falling onto it from a long way up, the air inside the bag escaped from some riveted air holes, providing a bouncy castle-like catch that was better than any modern crash pad, so long as you didn't miss it.

high point on a later attempt; a retreat might introduce pressure that interferes with future attempts. Nevertheless, I have always been grateful to that climber for making the call for me and speaking up. I guess it was probably obvious to him that he was going to prevent a ten-metre ground fall.

Immediately afterwards, I had what climbers would call a 'whitey', pale and freezing cold even though it was a warm summer day. Between my top-rope attempts on Obsession Fatale, Michael, our other climbing partner on the trip, had been top-roping an E6 which shared the same start and also had a very rounded crux right at the top. After my retreat, he announced he was up for a solo attempt on the E6. This time, I was part of the spotting team at the base. Michael climbed well to the last move, and as I stood below, arms in the air to spot him, I processed what had just happened to me. Completely distracted, I soon snapped back into the present when I heard Michael's foot slipping on the last move.

Next thing he was flying through the air, one foot scraping down the scrittly gritstone, eyes bulging with shock. It struck me how pointless it is to spot a falling adult human from ten metres. The lower half of the route had a complex series of rounded ramps and ledges, with which Michael was going to collide imminently and tumble in god knows what direction. Just as that thought passed, Michael hit the first ledge with both feet and absorbed the impact with a perfect gymnast's landing. Although it was a bit of a violent thud, he didn't fall down the lower part of the cliff. It looked almost unreal. Michael raised his hands and shrugged as if to ask, 'Was that it?' The older climber must have been dismayed by our attempts to win a Darwin award and walked into the forest, shaking his head.

The next day, Niall tried to onsight a big E6 called Barriers in Time. I fielded several of his falls from the last move, diving down the hillside to take in enough rope so that he stopped just a couple of feet above the ground. Niall just didn't seem fussed. It's hard to be objective about these things, but it felt like we were all influencing each other to get a bit too comfortable soloing with no margin, falling off and skimming the ground and pushing forward when we hadn't much idea if we were ready.

The same day, I top-roped another unprotected E8 called Doug, named after a man who lived in a small howff right below the climb. After my fright on Obsession Fatale, I had no real intention of soloing it; I just wanted to see if I could learn the moves well enough to feel solid and consistent on it. That's exactly what happened. Feeling good, I climbed it cleanly six times in a row, and it seemed like a waste of opportunity not to get my first E8. So I pulled the rope, laid my small bouldering mat out beneath the line and started climbing. It all went perfectly, and this time I was in a much better

Right: On Byres Road in Glasgow after breaking my ankle in the Peak District in 1997. © Claire MacLeod

mindset, calm and almost robotic in my movements, with no fear. As I rocked over through the crux, I thought: *This is more like it.* Just then, the tiny pebble I was pulling on with one finger broke off and I was falling. My left foot landed on the very edge of the mat, which in turn was sitting on a tree root. My ankle rolled over with a horrible crunch. The next week was spent in an orthopaedic ward in Stoke-on-Trent Hospital. My ankle has never been able to dorsiflex (heel drop) since, and it still gives me pain and limitation to this day.

It is difficult to judge the importance of this period in my apprenticeship as a climber. Is it really possible to have the perfect apprenticeship in managing risk on hard routes where falling could have disastrous consequences? Three months on crutches and daily pain thereafter was a powerful lingering reminder that no matter how much I wanted to push myself, I had to respect the hardness of the ground. I also appreciated that being able to do the moves on climbs was nowhere near the same as being a safe leader. There are many layers of judgement of rock quality, conditions, mental state and logistics that sit alongside the ability to crank on the holds. Perhaps if I'd taken a gentler approach to progression and hadn't gone through the pain and exclusion from climbing that followed, I'd have had a more serious accident down the line. I can't know. But I do think the episode was a catalyst for making me a safer climber.

Niall continued to make many 'skin of the teeth' leads on bold onsights, his favoured climbing discipline. Later, I belayed him on an E6 at Reiff in the north-west corner of Scotland. I'd onsighted the route myself and knew exactly what the protection was and how hard the moves were. Niall soloed up to a horizontal break with a single good hold. In the back of this are the sole runners on the pitch: three small cams and a skyhook, which, together, make a nest of gear just adequate for protecting the hard crimping above. Niall locked off to peer into the slot at the back of the hold, placed the first cam and gave it a tug. It ripped out and whacked him painfully in the face, which flustered him. He tried again and did exactly the same thing. Each of the small cams has a particular spot where they'll bite, and it is necessary to lock off in a strenuous position to fiddle them into place. Unable to get any of the cams to seat, in frustration Niall threw them over his shoulder onto the rock platform several metres below, then positioned the skyhook on the edge of the hold. He had trailed a third rope, clipped to the hook, that I would quickly stand on to tension the hook and reduce the chance it would ping off in a fall. When he reached up to take the first of the crux crimps above, I assumed he knew he had not yet clipped his sole runner and was just assessing the potential sequence above. Trad onsighting is like this, requiring multitasking of resting, arranging gear and assessing the upcoming holds all at the same time. He would surely drop back down, rest a bit more and swap hands on the hold so he could clip his own rope to the skyhook. But I watched in horror as he took the crimp and stood up. He looked tired from wrestling with the gear and wobbled onwards, quite possibly about to fall straight onto flat rocks at the base. Instead, he pulled over the top and peered back over with a big smile, laughing about throwing away the cams. 'Niall, you didn't clip the hook; you've just soloed the route!' He seemed mildly shocked for a few seconds, but five minutes later he started up another E6.

By the time the summer heat of 2001 faded into cool September mornings, I was more prepared to get on the lead on the headwall project above Chemin de Fer at Dumbarton. The summer of climbing lots of E6s had exposed me to many hard moves well above protection, as well as taking plenty of falls and fielding those of my climbing partners. I still didn't feel like an E9 climber, but I was ready to try and to fail.

Halfway up the E5 crack of Chemin de Fer there is a resting place at a good hand jam. I remembered being there the previous year, pumped and shaking. Now, on my first lead attempt of the project, I had the advantage of knowing the moves. Nonetheless, I couldn't help but notice the complete absence of any fatigue as I hung on the jam, staring up at the smooth headwall above. I have always expected climbing to feel desperately hard. To me, failure and struggle are what define it as hard: I have never revelled in the sensation of a move feeling easy, which seems to represent unrealised potential, coasting along inside my comfort zone. Hanging on the jam, my breathing was soft and relaxed, my arms warmed up but full of energy. I looked up at the iron road of old pegs lining the crack, and it felt time to leave its security behind and blast up the headwall above.

Often on rehearsed trad leads like this, it is normal to feel slightly detached from your body. All your physical effort and concentration are required to continue making moves, but a small space in your mind is able to step back and view such an intense reality from the 'outside'. For these two processes to continue, they have to operate in complete separation. The moment one process interferes with the other, focus can evaporate, and a fall is almost inevitable. So it was with some disbelief that I watched myself get through the first crux move and reach out right to the 'dead end' hold in the middle of the wall from which you are forced back left. With a cold gale roaring off the Clyde, the friction was excellent, and the spin move - reaching left from the sidepull to a crescent-shaped edge marking the final hard move - went perfectly.

The crescent is a lovely hold. Its lower half is sloping, the upper half more incut. It is necessary to take the sloping part first, and with chest pressed against the wall, bring your right hand over your head and carefully across your body to match on the incut part. If you do it too quickly, or let your centre of gravity drift even a few millimetres away from the wall, you peel off before the right hand gets there. I arrived at the move still feeling fresh, and brought my hand steadily overhead. The thought of imminent success crossed my mind; I'd never fall once the right hand made it to the incut. At that moment, the bony lump on the inside of my elbow skimmed the rock. This tiniest of bumps was enough to send me flying, staring down into the space below with the yellow lead ropes whipping in big arcs below my feet. The fall was huge, but it wasn't dangerous.

I felt a bit robbed of the ascent but had enough experience to know that on such a hard climb, even my tiny error of anticipating success was enough to lead to failure. In the two weeks that followed, I learned that my first attempt had been a better performance than I had realised. It's easy to assume that if you make one small mistake on a single attempt, you can improve on that incrementally each time, but if the climb is truly at your limit, it's not that simple. In my next ten lead attempts, I never got back to that last move and repeatedly fell off one or the other of the two cruxes just before it.

Cubby, who had been getting really into climbing photography, came along and abseiled down the wall beside me, taking shots during my attempts. It was fantastic to share the experience of trying to open a hard new route at Dumbarton Rock with Cubby, who had such a strong and similar personal relationship with this cliff. Requiem had been Scotland's first E7, and this would be Scotland's first E9 if I could finish it. But it didn't seem to matter how determined I was to fight my way to the top of the climb. I'd have my three lead attempts and fall off each one, the rope spiralling as I dropped away from the wall and took the ride over and over again.

One morning I asked Claire to come and belay me after the first chilly night of the autumn. We arrived first thing to find the crag looking very atmospheric, with wispy banks of mist floating off the Clyde and a piper practising his tunes on the shore below. There was a strong positive feeling that this would be a great moment to do the route, and I resolved to leave absolutely nothing on the table in terms of effort. Still, I fell and fell. Forcing it didn't seem to make any impact, at least on such technical climbing as this.

The following week, university reconvened - the start of my final year. On the first day we received a spirited pep talk from our course leader, who reminded us that the first three years of our degree had been a lot of fun, with plenty of time for various activities outside of study. But this year, we'd have to focus. He'd seen a long line of students fail their finals because they didn't focus when it really mattered. I took this all rather seriously and decided I couldn't fully immerse myself in climbing and training as well as prepare for exams. The pressure to return to Dumbarton lifted a bit, and I had a week with just a couple of training sessions at the wall.

Joanna George rang me for some details about the new routes I'd done over the summer for the column she wrote in a climbing magazine. Since Cubby had been taking pictures of me the previous week on the project, she asked about it, and I explained how disappointed I was in myself for not having figured out how to make one more crucial move of progress, and how I might not be fit enough to keep trying since university would likely be much harder this year. I told her that, in hindsight, I could think of countless little things I'd not done out of laziness or lack of thorough preparation. I could have squeezed in another attempt or training session here and there. On some of the attempts, I questioned whether I'd let the anticipation of a likely fall take the edge off my commitment as I slapped for the crux holds. I asked Jo what she thought I could do more of, or differently, to get the project over the line. I really respected Jo's knowledge and experience - she was a brilliant climber herself and had spent time around many of the best climbers in the UK, if not the world. She said that she didn't really know, but one thing she was certain of was that it was not for lack of effort and that I was one of the most motivated and determined climbers she'd ever met.

Jo's definitive reply left me taken aback, and I spent a lot of time afterwards churning over what it meant. I was pretty sure she wasn't just trying to be kind, but instead had given me a clear message that if I had room for improvement, perhaps I was looking in the wrong place.

Jo had also said that she would be in Glasgow in a few days' time and had asked if I wanted to join her for a session at Dumbarton. So I met her there one evening, and she offered to hold my ropes for an attempt on the project. After my light week of training, I felt I'd have lost the shape I'd need to have a worthwhile attempt. In any case, I'd need a long warm-up, so I suggested a tour of the boulder circuit first. We bouldered for an hour, and it was very light-hearted, just chatting and discussing the moves on climbs that felt easy. I couldn't help notice how different this was from my usual sessions alone, which were intense with much more time on moves I couldn't yet do. A front was coming in, and the October sky suddenly darkened, with rain imminent. Jo asked again if I wanted to get on the project. I knew I had to go now, or not at all, and couldn't pass up the chance to at least try. To be honest, I expected it would probably be raining by the time I'd got the ropes out.

Back at the hand jam rest on Chemin de Fer, I had that same feeling of effortlessness.

The wind was increasing by the minute and, with the leaden sky, lent the atmosphere an intensity I was more used to - a bitter, acquired taste of intimidation. It dominated the mood of the moment, and every gust seemed to ratchet up my readiness to attack the crux. On the headwall I was only vaguely aware of the odd thing: a sharp intake of breath, my shoe scuffing up the wall on the first crux and Jo's distant shouts of encouragement in her soft Highland accent. Things came into focus again with my right hand arcing over my head, homing in on the good part of the crescent hold. I felt the hairs on my forearm skim the surface of the rock. My fingers latched the hold and I fought to the next one right at my limit, tiredness and self-consciousness creeping in. Somewhere in these last few moves, I'd cross an invisible line between a big, exciting fall and a devastating one, if I didn't stay on. The next hold felt bigger, and I could still pull on it. The next one was bigger again; I could breathe and then pull. Then finally, a massive flake and I could stop and scream. Through the gale and spits of rain, I heard Jo's voice from below shouting, 'Brilliant!' It did feel brilliant. The holds beyond felt so secure and huge, the contrast with the tiny sloping edges below so stark. Big blobs of rain splashed onto the chalked jugs as I pulled through the easy finishing moves.

Back at the base of the route, the heavens opened. I said goodbye and thank you to Jo, who had to drive north, and I headed off to meet friends from uni for a session at the wall, followed by a long night in a Glasgow nightclub. Maybe I could survive the final year of my course after all.

I called the route Achemine, French for 'to move forwards' - the first route to venture away from the 'iron road' cracks and push a line up the face between them. Yet it only took one little corner of the wall and not even the highest part. The story of that wall was far from over.

Walking home through Glasgow at 3am the following morning, and for weeks afterwards, I spent a lot of time reflecting on how this process had played out. Climbing E9 was one of the two lifetime goals I'd set myself, and I'd managed it ahead of climbers in Scotland who had much more raw ability. I was desperate to resolve an accurate picture of how this happened, lest the value of it be lost. It was a confused picture. Jo's words about being the most driven climber she'd met rang in my ears, but I still struggled to believe this was true. She was, after all, married to Cubby, who always struck me as being utterly dedicated to rock climbing. Motivation and passion may be applied in subtly different ways, though.

If Jo's comment was true, my perception of myself must be pretty skewed, and I had an opportunity to learn something useful. I'd never reach my potential if I failed

Left: The first crux move of Achemine (E9 7a), kicking the left foot out to reach rightwards to the sidepull in the middle of the wall. The crescent hold can be seen in the top left of the frame; the lower tick mark for the left hand and the higher tick for the right hand to match on the incut part of the hold. Years later, Barbara Zangerl found a more direct way to climb the headwall, staying further left by using an undercut at the base of the zigzag score in the rock. This misses out the whole crux section, which remains unrepeated in 2024.
© Cubby Images

to recognise my key strengths early enough to learn to wield them properly. At first, I felt I hadn't taken the process all that seriously, berating myself for what I perceived as lack of forcefulness or initiative to make progress in my preparation. Later, I could see I was in fact highly focused and self-critical when it came to to pushing forward with the project, and since I was constantly on the lookout for errors or lapses in my commitment, it was difficult for me to see when I was bumping up against diminishing returns. I'd tied my satisfaction with the quality of each attempt very closely to physical effort and disinhibition in the moment - things I knew I could control. However, my first attempt, almost succeeding, and the final successful lead had one thing in common: a cold gale force wind. I couldn't control the Scottish weather, but I might have had the insight to recognise that I was only a northwesterly front away from success and to relax a little, given that October will always provide it.

At no point did I have any lingering emotional discomfort after failed attempts on the climb, and I took this as a reassuring sign that I could endure, even enjoy, sustained battles on future climbs. This aspect of sport has always felt quite straightforward to me. I want to see progress, but acute failures don't tend to bleed into a deeper emotional reaction, perhaps because of the simple truth that they are entirely expected if the project is genuinely hard enough. The freedom from a need to identify with the outcome of any single attempt had remained with me since my first day at Dumbarton Rock eight years previously, when I'd watched those climbers struggle on Requiem. Where it had made the idea of rock climbing attractive to me as a kid who lacked confidence, it now gave me the confidence to interrogate each failure and find a path forward. Only failure to gain some sort of insight, a new angle of attack, from each attempt bothered me. A route completed is an aesthetic creation in its own right, but as a sporting achievement it is just the platform for what really matters: moving the needle on what is possible.

Marching through Glasgow in the middle of the night on the way home from the nightclub, my mind was racing. But not with thoughts of celebration. I reminded myself I'd just climbed my own and Scotland's first E9, but I couldn't maintain focus on that thought for more than a few seconds at a time. It had been my lifetime goal but no longer seemed to matter, even a few hours after achieving it. I'd noticed this feeling before when completing hard climbs and had even felt guilty about it. As my pace quickened along Great Western Road, I finally let go of a sense that I should dwell on a goal that was now in the past. I came to understand that the way I celebrated completing a climb was by applying the learning to the next one. Celebrations could begin on a steep wall near Arrochar the next day.

10.

THE FUGUE

Left: Claire using a remote shutter release to take a photo to use on our wedding invite in October 2001. Behind is the Upper Crag in Glen Croe and the Fugue (E9 6c), which I climbed later that day.
© Claire MacLeod

After Achemine, my mind was immediately drawn to a couple of other new lines I'd previously looked at but was not yet ready for. I wasn't bothered all that much if others did them before me. I was quite conditioned to feel that they might, since many other climbers who put up new routes seemed to worry and talk about this a lot. To be honest, I expected that other climbers could do routes quicker than me, if they took the notion to try. So what was the point in worrying? All I could control were my own attempts.

Many of the 'rules' in climbing are unwritten traditions that survive across generations. One of these is that the effort of finding, cleaning and attempting a new climb is respected. If it is common knowledge that if someone has found a potential new route and is either actively trying it or intends to return to it after an enforced break due to conditions, injury and the like, they are given time and space by others to complete the project. Nevertheless, just as in other sports, good sportsmanship isn't always maintained, and history is littered with stories of climbers being beaten to the first ascent of their own projects.

At times, this is so blatant that it might be fairly called 'stealing' someone else's vision. Usually, the person doing the stealing comes up with some sort of ad hoc justification for it, which often makes more sense to them than it does to those looking on. Sometimes it does make sense. For example, some unclimbed lines are so glaringly obvious that it becomes awkward for an individual to 'claim' that piece of rock and for others to be forced to leave it alone while they try it. Some rock features also become well known and openly discussed, or are mentioned in guidebooks as a 'last great problem', an obvious line that stands out on the crag as not having been done, usually because it is hard. I desperately wanted to avoid confrontations over new lines, yet early on in my experience of new routing, as I began exploring the big

Right: With Richard McGhee in Glen Nevis.
© Cubby Images

crags of the Highlands and Islands, I did become involved in such a controversy.

Arran is a beautiful island in the Clyde estuary and its skyline of granite mountains dominates the outlook from the Ayrshire coast. Those mountains contain a huge amount of routes concentrated in a small area and are on the 'must visit' list for any climber from Glasgow. Flicking through the guidebook, I noticed the reference to a 'last great problem' on a stunning but rarely visited cliff called Cioch na h-Oighe. The topo drawing in the guide made the unclimbed line seem all the more impressive. Essentially all the routes bent in crescent-shaped detours around a massive central bastion of rock with no lines going through it.

A climbing partner from uni, who came from Ayrshire and had climbed many times on Arran, encouraged me to go there with him. So a team of us took the ferry over, and we had a fantastic trip. Everything about the place appealed to me. Walking in to Cioch na h-Oighe for the first time, I thought the corrie looked beautiful. The crag was imposing, but the place didn't feel serious or intimidating. On the first day there,

I onsighted the hardest route on the cliff, a bold E6 first climbed by Kevin Howett. It was one of my first E6 onsights and I felt quite comfortable on it and found the granite lovely rock to climb.

After this, I abseiled down the unclimbed central wall for a look. The route would be three distinct pitches. Pitch two would almost certainly form the meat of the difficulties, but the top pitch might also be tricky, with a few options for where you could go.

After abseiling to a big ledge at the top of the main pitch, I re-rigged the rope and continued down. As soon as I was over the lip, I was surprised to find that the whole line was not only obvious but very well chalked. I had enough experience to know that the route had seen attention from someone who really knew what they were doing. Each hold had been carefully cleaned, and every single foot smear had been marked with a tiny dab of chalk. I'd seen that done here and there on boulder problems before, but never for every foothold on a 45-metre mountain pitch like this. I was anxious and unsure what to do. There was no way of knowing who was trying the route, or if it had been completed already. My hunch was that it probably had been done, as the pitch looked like it had been prepared for a lead. All the same, I didn't want to go ahead and climb it. But I was intent on at least trying the moves, and I found I could do them.

On the way back up the rope, I also tried the top pitch. Unlike the one below, it had no chalk, and some of the holds were covered in thick, tough lichen. I spent about four hours scrubbing the whole wall with a wire brush to make the moves possible for me. It was clear this pitch was only a grade or so easier than the one below.

Back home in Glasgow, I made an assumption. Even if the main pitch had been done, whoever had done it had gone a different way for the top pitch since that was still dirty. Most likely, they would have taken the path of least resistance, escaping left after the big ledge. A month passed and I heard nothing in the climbing media or on the grapevine about the route having been done, so I resolved to get on the line I had cleaned. Regardless of whether the main pitch had been climbed or not, the headwall would add distinction and a lot of difficulty to the overall route.

On an optimistic forecast, I returned to Arran with Richard McGhee, a friend from uni who was becoming my main climbing partner. Richard is a great character, a polarised mix of fun-loving yet serious and conscientious when it matters. We had spent many weekends around Scotland, as well as England's Peak District, and with Richard it was mandatory to seek out a local nightclub, no matter how small or quiet the town we were staying in. I usually, but certainly not always, paced myself carefully with drinks to make climbing on the Sunday possible. Richard didn't drink a drop of

Left: Falling off the last move of Kaluza Klein (E7 6c) at Robin Hood's Stride, Peak District. Richard is turning to jump off the ledge and take in rope as I fall, which is essential to stop the leader hitting the ground. My ankle caught the rope, flipping me upside down, and one of the two cams ripped. I bounced off the ground on my back on the stretch of the rope but got away with only a headache.

alcohol until his 18th birthday and, perhaps because of this, he went through his hard partying years later than most of us. By 19, I'd done enough of that, and the novelty was finally wearing off, allowing me to get on with my life.

On the cliff, Richard was determined and careful, sometimes to a fault. It's common when placing protection in cracks to give cams and nuts a stiff tug, both to seat them so they are less likely to fall out and to test that they might hold a fall. But there's a tug, and there's a tug. It was quite disconcerting to watch. Once, while on a multi-pitch route on Buachaille Etive Mòr in Glen Coe, Richard had led a long pitch and arranged a belay anchor on a ledge. He placed a cam, tugged it hard, and when it popped out, he lost his balance and fell off the ledge, taking a 100-foot head-first dive down the cliff, past the previous belay.

Our trip to Arran tested both our reserves of motivation. The rain was incessant, our remote camping spot in Glen Sannox had no nightclubs or even a pub to retreat to, and for four days we trudged around ridges and wandered around the foot of soaking crags. Even walking uphill I couldn't seem to get warm in my wet clothes.

Finally we woke to a break in the rain and headed for Cioch na h-Oighe. The climbing on the main pitch of the new line would be too hard for Richard to follow, so we planned that he would second the first and third pitches and jumar up the rope on the second. The first pitch was still quite wet, and I struggled through it. Richard slipped off wet starting holds, swinging off into space, and was forced to climb the rope on this pitch as well. Meanwhile, the sky darkened and the wind strengthened to a loud gale, the next weather front well on the way. In the intense atmosphere, I dispatched the main pitch, gritting my teeth at the end as I wobbled with numb toes on delicate smears.

After an exposed trip up the rope, swinging in the gusts, Richard joined me. The wind was now so extreme we could barely communicate even standing right next to each other, and the situation felt a little intimidating. With a barely-protected section of climbing above the large platform, I contemplated the situation if things went wrong. Were I to fall and break bones on the ledge, staying here for several hours in this wind wearing just a lightweight top could well end up being fatal. However, if we retreated, I'd only be back and have to lead the bold pitch below a second time. I weighed it up and decided to get going before the sense of momentum had a chance to fade. My chattering teeth reminded me there would shortly be no decision to make if I didn't move. I signalled to Richard with an upward point and a thumbs up, and set off.

The roaring white noise of the wind, the sight of Richard getting blasted on the belay below and the ropes billowing sideways killed off any impulse to hesitate. As I pulled over the final bulge, the turfy ledges of the exit slopes came into view, with just

Right: On the Isle of Arran.
© Richard McGhee

some delicate slabs to go. But big splashes of rain now arrived on the wind; the front was about to hit. When a front arrives in the mountains, a cliff can go from completely dry to pouring with water in seconds. There wasn't time to stop even for a moment and worry about this; I could only accelerate up the slab in one continuous motion as it turned from pale grey to dark red in less than a minute. Richard followed, climbing the rope as swathes of water streamed down the sodden cliff. I tried my best to contain my delight at having pulled off the ascent with absolutely no time to spare, since Richard's climbing trip had consisted of about five metres of actual climbing, five days being wet and cold and zero time in the pub.

    A month later, it emerged that the climber who'd been on the route was John Dunne. My heart sank. John was one of the best climbers in the UK in the 1990s, and he was quite an unusual character. A huge, bulky Yorkshireman more suited to rugby or strongman competitions than hard rock climbing, he also seemed to have a personality to match, with a reputation for being quite confrontational. Every famous hard climb he'd done seemed to have one controversial story or another attached to it. The last thing I wanted was to be involved in one myself.

I'd only met John once before, when I'd gone for the first time to the Kendal Mountain Festival in the Lake District with my friend Nick from uni. Nick had made a really nice film about our university mountaineering club and was showing it at the festival, which for decades has been known as a big gathering for climbers from all over the UK. As part of it, there was a bouldering competition in the local climbing wall, and for something to do during the afternoon, I entered. Since I have always struggled with indoor climbing compared to outdoor rock, I had only taken part in a handful of competitions, soon after I started climbing. One Saturday morning, when I was 16, I was waiting for the train to go to a local competition in Glasgow. As I sat taking in the bright, sunny morning, it suddenly seemed crazy to go to an indoor competition instead of a crag. I got up, walked over the bridge and caught the train going in the other direction, to Dumbarton Rock. After that, I never took competitions seriously and always wanted outdoor rock climbing to be my focus.

So the Kendal comp had been my first in a couple of years. As I finished the last problem and handed over my score sheet, the organiser glanced at it and immediately told me I'd got into the final. At the comp that day, I'd seen many of Britain's best climbers for the first time, so it was a nice, if awkward, surprise that I'd made the next day's final of the top seven, ahead of John Dunne. I got even more of a surprise when John walked over to me and asked, 'Are you David MacLeod?' He sounded quite gruff, but that was his reputation, after all. 'I saw you repeated that route at Gorple and thought it was E7. That was pretty funny,' he said sarcastically. John was referring to an E9 called Carmen Picasso on a quiet and obscure crag on the Yorkshire gritstone edges, which he'd put up the previous year. The route featured in a recently released DVD about John's climbing, which I had eagerly bought and watched. He was not only doing hard and bold routes on gritstone that I aspired to, but climbing on mountain crags as well. This, together with John's story of being isolated from the mainstream climbing scene while living in Yorkshire and then Belfast, made me relate to him, despite his self-confident demeanour. The footage of him on Carmen Picasso looked great, and I thought that a short, intense route like this might suit me.

So, during a weekend on Yorkshire grit with Richard, I had asked if we could go and look at the E9 late on the Sunday, with a view to trying it more on the next trip. I abseiled down it and found a very poor micro wire placement, which John hadn't used but I felt had at least a chance of holding. I also managed to do the moves straight away. Temptation got the better of me, and as the light faded, I tried to lead it. I shouldn't have: it was too dark to properly see the footholds, and time ran out on tired arms as my fingers crept on scrittly slopers at the crux. I fell off and the runner held, but one of the two strands of wire had rolled over the grit as it came tight and snapped. Somehow, it had become wrapped in the other strand and cinched up on it, holding my fall. I was very lucky to avoid hitting the ground. With a new micro wire and fresh arms, I dispatched the route without incident on my next trip.

Grading controversies are one of the biggest headaches in new routing. Although entirely subjective, it still feels a bit confrontational to downgrade someone's route. It's definitely true that climbers have done this as an ego trip, or simply because they don't realise how well they are going at the time. It's something I'd rather avoid, but agreeing with a grade when you strongly suspect it is incorrect, even just implicitly by saying nothing, can feel awkward, even dishonest. For me, Carmen Picasso felt much too easy to be E9, and I was sure it was no harder than E7. A two-grade drop is particularly harsh, but I felt that if I didn't give my honest opinion, I was complicit in inflating the route's grade, essentially lying by omission. The lesser of two evils was just to call it as I saw it.

I began to question whether it really was the lesser of two evils as John towered over me at the Kendal comp, powerfully explaining why I was wrong and why I should reconsider both the grade and my attitude. He was clearly not impressed and others in the busy climbing wall noticed the confrontation, as perhaps was John's intention. I was mortified and just wanted him to go away. Over John's shoulder, Nick was sniggering and making faces at me, and I was sure that if my expression cracked and I laughed in John's face, there was a good chance I'd have my own rearranged. I held it together - at least an E8 effort on my part - and John marched off.

These kinds of disagreements rumble away continuously with hard trad routes, especially on second ascents. After that, the grade usually settles out. But the uncertainty can really get in the way of shared enthusiasm for the same piece of rock and climbing experience. I find that the best approach is to keep in mind that everyone's subjective judgement of difficulty is vulnerable to bias, and the experience of climbing a route is never the same for two different people. I try to accept that grades of new routes will change, often downwards, but it is better in the long run that climbers freely give their honest opinion. That seems to get grades to the settling point as quickly as possible, although it is easier said than done.

Given that tense first meeting with John at Kendal, I was dismayed to learn that the route on Arran was his. After some confusion, I heard John had indeed climbed the exact same line as me, including the final pitch. I doubted this could be correct, since it was so dirty and the main pitch so thoroughly cleaned. But John rang me and asked me to describe in detail where I went, and he told me that was what he had done. He said the top pitch had been dirty, but he had just climbed through it. Thankfully, there was nothing more to it, and he has always been friendly with me ever since. John called the climb The Great Escape. Had the first ascent been mine, I would have called it Macrochiera, after the giant spider crab, since the wall resembled the shape of the crab, with the flake of the crux pitch forming the left pincer. The right pincer is a much harder line and remains unclimbed to this day.

Although John could be a little abrasive and hard-nosed when it came to the game of being a professional athlete, I still found his climbing impressive and interesting. He, like me, had a significant issue with being lean enough for the elite end of the sport, but he could make up for this with very precise footwork and by dropping excess weight to get in shape for hard projects at the right moment. His lack of form when two stone overweight made him an easy target for critics. I've noticed many times since that some naturally lean elite climbers simply don't grasp the scale of fluctuations in ability among those who experience large swings in excess fat levels.

After this whole episode, I briefly wondered if I too should worry about being beaten to new lines, and whether I should approach them with more urgency, but the funny thing was, it almost never happened to me again for the next 20 years. In fact, the only time since that I've been beaten to a first ascent that was really important to me was on a boulder project in 2022. Doing difficult new routes in the mountains takes a lot of effort; weeks 'wasted' sitting in a tent waiting for a break in the rain and days cleaning dirty rock and trying moves. A sense of urgency forces you to get out on a shaky forecast and make progress on big projects where it would be easy to opt for more accessible crags. Nonetheless, in later years I gradually came to realise that the likelihood of someone else being motivated enough to climb the routes I wanted to do was pretty slim, and I quickly stopped thinking about it at all. Moreover, with the advent of social media, it's now a lot easier to let people know that you are trying a line, and it is still seen as poor form if you 'steal' someone's project if they are actively trying it.

Rather than competing for hard new routes, collaborating on them with others appealed to me more. I had still never shared attempts on any climb harder than E7 with someone else. One of the best things about rock climbing is that the very hardest routes are often right next to much easier climbs, both at the crag and indoor walls. If you reach a high level in the sport, you don't have to isolate yourself from the friends and partners you started climbing with. I was grateful for this, especially since connecting with other climbers I knew by reputation still wasn't something I felt confident enough to do. Since my trad climbing was still progressing, this didn't bother me as much as it should have. With hindsight, I was primed to be heavily influenced by the first partnership I might make with someone keen for the same routes as me. Jo George's assessment of my struggles with Achemine still hung in the background of my thoughts as a hole in my understanding of my own climbing. Staring into the gap was not helping. I needed someone else to show me what was there.

Life as a climber, at any level, sets a person up to become familiar with and then comfortable doing things differently from most of society. When forecasters warn of winter storms and icy roads and urge us to stay indoors, we frantically sharpen ice tools and pack rucksacks. Rejection of the notion that things should be made as safe as possible, such as by placing bolts in the mountains, runs through the entire spectrum of Scottish climbing culture. Those who can see the value in this way of thinking become lifelong participants, reinforcing the bias among those who climb. Nonetheless, elite climbers appear to have another 'gear' that even their peers simply do not; an ability to tolerate, if not revel in, extreme physical and psychological rigours on the cliffs. I wondered if I might be able to access this gear, even just once in my lifetime. The problem was, although I had seen the results of those operating in it, reflected in the routes they created and the confidence they projected, I'd still never witnessed a climber finding this gear and therefore struggled to define it. As I understood more about the logistics of hard routes by observing the best climbers from afar, I increasingly felt that these practicalities could not account for why they stood out. I was looking at effects, not causes.

My formal study of sports science, nearing its completion, reinforced this view. Adjustments of physical training cycles and psychological practice were important in elite athletes, but only in prising open small gaps between those who were already very different from most other people, even in their sport. A magic ingredient, if there was one, still seemed to be more about who they were rather than what they did. Perhaps surprisingly, sports science seemed rarely to engage with the obvious question: can this be trained? And if so, how? Professional sport tends to wait for new talent to emerge, already carrying the 'disease' that could make them withstand the onslaught of hard training, relentless competition, bitter failures and even danger. The primary job of coaches and sport scientists was the final shaping of the professional athlete, often rounding off the hard edges of fierce personalities, even restraining them from driving themselves into the ground, by programming in a compulsory rest day or a family event. The very best athletes seemed to me like rockets, with coaches acting like tiny thrusters on the sides, tweaking the direction of the raging force driving the machine forward. Ironically, sports science stared more directly into the fire inside athletes, when their unreasonable tendencies spilled over into extreme behaviour that negatively affected performance, such as eating disorders or performance-enhancing drug abuse. Yet even in this area, it struck me that little had been achieved in containing the damage athletes could do to themselves if they veered off course. Reading between the lines of the eating disorder literature, I sensed the frustration of the field in trying to influence it. It was the dark side of the same coin of elite performance.

My studies had so far shown me little to answer my question of whether this extra gear could be accessed with training. I needed to look differently at other climbers who could reach this gear, not just by climbing their routes but by sharing their rope.

It was outside of Scotland, on the geographically tighter circuit of gritstone crags in the Peak District, where these connections were formed. During the dark winter months, Richard and I continued to hang out there at any opportunity. The Peak didn't suffer the continuous rain and storms of December in the west of Scotland, but there were long periods of damp, foggy weather in which we nursed hangovers by wandering around the gritstone edges, lurking below the E8s and E9s put up by Johnny Dawes, Jerry Moffatt, Seb Grieve and the other famous climbers I'd watched in the movie *Hard Grit*. The Peak was a top climbing destination, but in the December drizzle, the crags were eerily quiet. Yet seemingly every time I found myself under a dripping E8 waiting for the clag to clear, the same awkward-looking youth would appear out of the mist, staring up at the same climb. Tom de Gay's obvious enthusiasm and knowledge of these climbs overtook his apparent shyness, and after several such coincidental meetings, we found ourselves climbing together.

The same weekend I'd tried John Dunne's Yorkshire E9, I'd bumped into Tom at Ilkley crag. I'd onsighted a short E6/7 called Desperate Dan and was keen to look at another E7 called Deathwatch. Leo Houlding, another top English climber, had onsighted it before, albeit at the height of his teenage fearless phase, and it sounded doable. We arrived to find it well chalked up, which was a great advantage for a successful onsight. Tom got ready to jump on after me, but warned that he hadn't been climbing much and wasn't fit. He always seemed very relaxed about leading these bold routes and couldn't seem to make himself worry about the consequences of hitting the ground. Tom had climbed over 60 gritstone E7s, and he shocked me by saying that he'd fallen off a third of them. 'I should really stop falling off,' he said, in the same way someone might scold themselves for forgetting to put the milk back in the fridge. I was amazed that he'd escaped injury with that approach, especially given my three months on crutches after a six-metre deck-out at the Roaches. Perhaps young man's

Left: Onsighting Deathwatch (E7 6b) at Ilkley, Yorkshire. © Richard McGhee

logic deduced that so many falls without injury meant that the routes were more scary than actually dangerous.

Nevertheless, as he wobbled upwards for several metres on Deathwatch, he at least began to look scared and engaged reverse gear. The climb starts by moving out across a wall from a gully. I had opted to place a cam in the gully and lead it, but the runner was only really protection for the first move or two, thereafter providing the theoretical benefit that it might stop you rolling too far down the hill if you fell from higher up. Tom had opted to solo, so I tied myself to the runner and spotted him with the intention of guiding him onto the flattest part of the landing if he fell. As he struggled to downclimb, the reality of the situation hit, and my stomach turned at the prospect of watching him tumble down the gully. I leant outwards from the cam and stretched my arms out, visualising trying to catch him by the T-shirt as he plummeted past. I wasn't even kidding myself that this was realistic anymore. Tom made it back down the lower moves with a fight, and I breathed a sigh of relief. However, to my dismay, he had two more attempts immediately afterwards, both ending the same way but in a progressively more harrowing fashion. At least he was being true to his resolve to quit falling off. Although his climbing was unsettling to witness, and I worried that he could easily break himself, I greatly admired his commitment to pushing himself so close to his physical limit on bold climbs. I'd rarely seen someone so ready to do this and I liked the energy.

I'd been telling Tom about some great routes I'd done recently in Scotland, and he said he'd come up in the summer to take a look. In particular, I wanted to show him a beautiful E8 called Femme Fatale at Whale Rock in Glen Nevis, first climbed by Cubby. I'd done the second ascent and thought Tom would love it, but when we arrived at the crag in June, Tom pointed at a blank, off-vertical wall just to its left and asked, 'What's that?'

'Nothing,' I said. I wasn't just saying 'nothing' as in not a route. I meant 'nothing', as in blank. It just looked impossible to me. I could only see one hold on it, which was an upside-down finger flake at 15 feet. That was it.

Tom didn't seem to care. 'Let's get on that!' he said. I felt it was a waste of our first day, but I'd come to the crag to belay him on Femme Fatale, so it was up to him.

He hung a top rope down it, and after inspecting some tiny ripples could actually make a few moves. Watching him, my curiosity kicked in, and I wanted to try as well. The moves were ludicrously thin. Holding the finger flake as a weird, backhanded undercut, I could stretch up for a minuscule edge and bounce again for another at my full arm span. It was fascinating to piece a sequence together with Tom, but our progress stirred an ache of frustration, since I couldn't envision this beautiful series of movements being transformed into a complete route. Even if my judgement on this front was premature, the wall had not a single runner until the easy finishing moves. Moves as hard as this couldn't be justified as a solo, at least for me.

'That's definitely E9,' Tom said. He had put on a lot of muscle since the winter and his physical strength was rapidly catching up with his boldness. It was an impressive combination, and watching him on it made it seem physically possible; listening to him, even more so. 'You just need to try that some more,' he said. 'You'll do it for sure.'

I thought he was just being encouraging, but later, I wondered if perhaps he actually thought this was true.

Conviction certainly ran through Tom's own climbing, and I learnt a lot from it. The next day we looked at another new line - a stunning glacier-worn wall overlooking Steall meadows in an exposed but inspiring position. There was an obvious line right up the middle, previously climbed by Gary Latter, using a bolt to protect the crux. This was during a time when climbers in various countries were experimenting with bolts, and Gary must have felt the route wasn't possible without one. His decision ran counter to the ethic of strict separation of bolted and traditionally protected climbing, and the bolt had been chopped by Gary's contemporaries, leaving the route essentially defunct. We found a nest of very optimistic gear in a little patch of quartz at half height that, realistically, wouldn't hold a fall. The most comical placement was a sideways skyhook stuffed into a little hole and cammed in place. It would assuredly snap if a climber fell onto it. We laughed darkly at the prospect of getting away with a fall from the desperate crux slap. You'd hit the belay ledge ten metres below and roll off, with another drop onto the next ledge underneath to finish you off. We carried on the dark speculation: the rope would cut on that sharp ledge, you'd roll down the rest of the crags and drown in the river, then wash out to sea and be eaten by sharks. The point was: don't fall off this one.

There would be no prizes for a tentative approach. If either of us decided to lead, it would need to be done with total commitment and no hesitation on the slap. I tied in and led it only because I felt really at home on the steep wall on small crimps and was sure I had enough of a margin. To my relief, Tom, worried he wasn't fit enough and having already fallen off a couple of times on the top rope, decided against it. My ascent had felt harrowing enough; watching a friend's would be far worse. But the next morning, as we were due to leave, Tom said he'd like to go back up to it. On the lead he looked fine at first, but as he approached the crux slap, he was starting to fight. I could read the tension and hear the effort in his breathing, and I braced to witness a horrific fall as he launched for the slap. As he caught the hold, the rest of his body was already starting to drop. My stomach turned. Every muscle strained to the limit to resist as his trunk arced out from the wall. As his left arm straightened, with no more room to absorb the backswing, Tom let out a fierce growl through gritted teeth. He didn't let go.

It was one of the longest seconds I've experienced and considerably more stressful than my own ascent. Yet afterwards, Tom seemed only mildly perturbed. To him, it was quite straightforward. Recounting the joke about the sharks, he told me he just knew he must not let go of that hold. Simple as that. It was a lesson for me: when the shit hits the fan and climbing becomes a matter of survival, treat it as such and just don't let go. It is well known that an extra few per cent of strength exists behind a wall of inhibition to keep muscles and tendons safe. Perhaps to tap into this, you have to clearly understand and face up to the fact that your life really does depend on using it. Detaching from or minimising fear may be counterproductive.

We called the route Impulse and gave it E8. It seemed to nicely reflect the desperate crux slap that was hard to initiate from poor holds, as well as Tom's instinctive decision to go back up and lead it, and his quick reaction to turn an unfolding disaster into success and survival.

It had been a scary day out for me. I would have happily drunk tea and calmed down for a day or two, but Tom just wanted to get on the next project, so we took ourselves to the Arrochar area. The weather wasn't great for going up high, so we went to some great crags on the lower slopes of The Cobbler, where I knew of another futuristic project, which, like Impulse, had been bolted by another hero of bold trad climbing in Scotland, Paul 'Stork' Thorburn. I'd never climbed with Stork, but he seemed typical of many great trad climbers in Scotland at that time and since. He was immensely experienced and technically gifted, as well as having a very cool head and the ability to get himself out of every sticky situation on bold new routes. However, if a climber lacks pure bouldering strength, there is a hard limit to the difficulty of the climbs that are possible - if you cannot pull on the hold, technique will not work as a substitute. Even though Stork had bolted this line, he'd still failed to climb it, and the bolts had since been chopped.

Both Tom and I could do the moves. Again, Tom judged that the route was too tough for him to climb on this visit, but he encouraged me to keep trying. My experience with this style of climbing, working hard trad projects on a top rope first and then leading them, was growing, and I had developed a strategy. I'd base my judgement of when I was ready to lead on the consistency with which I could link the moves on the top rope first. Once I'd linked the route, I'd come back on 'lead day', warm up and top rope the route once. If I could link the moves first try, generally the lead would be on that day, before there was any time to burst the psychological bubble. However, on my previous top rope sessions on this project, I'd only ever linked the climb once in a session, usually on my first attempt. On the second, I'd tell myself, *Right, this time I'm imagining I'm leading.* That try, I'd fall off the crux move. I did that for six sessions in a row - not a good sign. Even if your belayer could dive down the hill fast enough to keep you off the ground, a lead fall from the crux would still end with a horrific smash into ledges at the foot of the route.

An obvious but potentially stupid deviation from my usual way of working emerged as an idea. For those six previous sessions, I'd linked the climb first try. So what about just turning up and leading it without a successful top rope first? It was a bit of a gamble, but two things led me to commit to it. First, I reasoned that six sessions had played out exactly the same way, which was at least consistent. Second, I had a hunch why the sessions were playing out this way. The crux was a grunt move - pure power to pull in on a tiny overhead undercut and violently slap to the next edge. On the second attempt of the day, as I was doing the move, I would try to understand the movement and execute it more efficiently. But this seemed counterproductive. It didn't yield to thinking; it yielded to grabbing and pulling like hell. On the first attempt of the day, I seemed to default more to this strategy. For better or worse, this was my justification.

On a chilly October morning, Claire drove me out to Arrochar and we approached the crag across frost-lined deep brown grass. Claire was studying fine art photography and she set up a camera on the slope below to take a portrait shot of both of us standing in the long grass. We'd use the shot for invitation cards to our wedding, which would be just before Christmas, assuming today went okay.

Claire knew the score, and we said little as I warmed up, arranged the gear and psyched up at the crag. By this point, Claire had already belayed more hard trad ascents of E8 or above, or at least attempts on them, than anyone else in Scotland. She had fielded most of my 70-foot falls from Achemine, and saved my life on the E4 nearby by taking in a crucial metre or three of rope as I plummeted to an otherwise certain ground fall. She had also spent enough time observing climbers to recognise the shift in atmosphere that occurs, or at least should occur, when the would-be leader realises it is time for the dangerous ascent to happen. Both of us had seen leads go wrong when climbers failed to take them seriously enough or, conversely, when tension morphed into anxiety and then to panic, when it was too late to reconsider. Claire had witnessed at close quarters almost my entire progression from a novice buying my first set of wires, to systematically working through the hardest trad lines in the country. What she had, and I did not, was both objectivity and the ability to read the person, not just the circumstances. When I came home from a day working on a hard route, she could tell the next session would be a lead before I even opened my mouth.

A climbing partner who grasps the subtlety of preparation for dangerous climbing and is not afraid to intervene is a priceless mentor. The key, which Claire understood perfectly, is to mentor at the right moment. She might not have known the practicalities of what was missing from my preparation for a dangerous piece of climbing: that I needed to try a different foothold on a given move, or simply build more of some specific aspect of fitness. But she knew, often before I did, that more work was needed. Likewise, she knew when the preparation was complete and it was time for execution. As I discovered holding Tom's rope on Impulse, belaying someone slapping for holds at their limit, where a fall would have horrific consequences, can be harrowing. Occasionally when belaying me, other climbing partners would fuss over small details right before my lead, perhaps an expression of their own fears about what was about to unfold. Claire had a knack of knowing how to remove obstacles for a leader, not least any fear she herself might have. I often found that desperate routes felt easier in her company. So much so that it became a running joke between us that all that had been needed was for her to come to the crag. Execution of dangerous climbs is about removing distraction, complication, stripping everything back to the quiet moment tying into the rope, ready for a fight to stay alive.

Just like the top rope attempts, I took the undercut, drew a deep breath and exploded with everything I had. I hit the edge and powered on to a jug on the lip of the overhang. Here, I shared a few words of relief with Claire as I rested. 'Save it for the belay,' she replied. In the silence that followed, I noticed that my ring finger felt sore on the edge of the jug. Was it just the sharp edge of the hold? Later I realised I'd pulled so hard on the undercut that I'd torn a ligament.

I called the route The Fugue. Scotland's second E9. My nan had introduced me to the idea of a fugue during a conversation about music. It is a format with a melodic theme running through it, and although it deviates and builds, it always returns to the core underlying theme. As my climbing had progressed beyond Dumbarton Rock to include many different places, styles of climbing and new routes, it had diversified hugely. I was still trying to grasp why I had arrived at this point, doing first ascents that were starting to break new ground in Scottish climbing, sometimes on routes that had already been attempted by people I greatly admired. When I looked forward to my next attempts on these climbs, or when I recalled what I'd enjoyed about them

Right: Great relief after successfully leading The Fugue (E9 6c) and a big contrast to the seriousness leading up to it. The relief was always short-lived and I was soon thinking about the line on Whale Rock that would become Holdfast.
© Claire MacLeod

Left: The headwall above the crux of The Fugue at Glen Croe.
© Cubby Images

after succeeding, it always came back to the individual moves, their intricacy and especially the process of solving them. It was while problem-solving on crux moves, either on a boulder or a futuristic hard trad route, that I'd get lost in focus and hours would go past. When the light or my strength faded and it was time to abseil off, I'd feel my whole body glow, burn and then ache with the physical effort I hadn't even noticed I'd been making.

I thought of the still-unclimbed project at Whale Rock. Watching Tom start to unlock the secrets of the moves, unearthing the possibility that this could one day make a complete sequence, had been one of the most exciting moments of the summer. Why? It wasn't even me climbing. I had little desire to celebrate routes completed. They were over; all the questions answered. The interest lay in moves yet unclimbed, unsolved. But even if I was repeating an existing climb, I'd usually find my own sequence. A better way, for my body at least. I was starting to see how important this insatiable fascination was in driving my climbing standard upwards. I wasn't that bothered whether moves or routes were possible in the end. After all, every time one

Left: High on Holdfast (E9 7a), locked off on the hold that gave the route its name. As I snatched at the hold, my foot would always slip a fraction of a second before I'd latch it. It was critical to grab it as fast as possible to avoid a ground fall.
© Cubby Images

route was shown to be possible, I'd just try a harder one. It was the finding out that was fun. Was there another way to use the footholds? Was there a ripple I'd previously discounted that could be used as a hold? If a method I tried didn't work, I didn't care. I'd just keep searching and enjoying doing so. Of course, the fact that I was unburdened with the need for it to work often facilitated a successful outcome. This approach seemed to be putting me in a strong position psychologically for grappling with hard climbs. Perhaps I could take this lever and push it a lot further?

In the deep frost of early December, I returned to the Whale Rock project and pieced together the whole desperate sequence. In the company of a friend from Edinburgh, Dave Redpath, I set up a lead day and linked the moves on the first try of the session. Dave had brought his video camera and he set it up on a tripod, but as I pulled the top rope and the mood turned serious, I noted that he never did turn it on. Later I asked him why. 'I don't do death watch,' he said. That was fair. This was also a route not to fall off. It would be a battle for survival, and I attacked the moves as if the climb were trying to kill me. On the last move, slapping for the first decent edge, my left foot would always slip right at the moment I grabbed the hold. It was critical to snap my fingers onto that edge faster than my foot would slip. After this move, I called the route Holdfast. As Tom had predicted, it was 'definitely E9'.

Starting up The Hurting (XI,11) in Coire an t-Sneachda, Cairn Gorm. At the time, launching up an open face in winter without either a crack or corner to aim for felt very intimidating.
© Steven Gordon

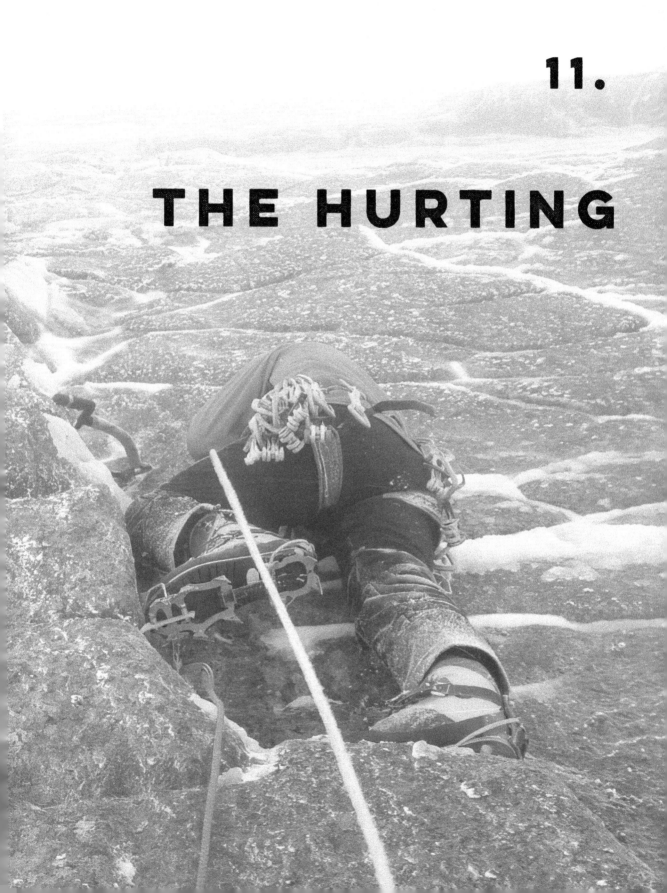

11.
THE HURTING

Claire and I were due to get married a week after climbing Holdfast. I'd now been with Claire for six years, and I found time with her replenishing. She is a fun-loving but introverted person, and being in her company was a light-hearted contrast to the relentlessness of pushing myself in climbing. I enjoy that relentlessness more than is good for me, and to this day, when we are apart, I find myself overcooking all aspects of my life: training too much, working too much, sleeping too late.

Despite Claire's deeply positive influence on my life, getting married at a young age and thinking about how my life might pan out scared me and felt like a big deal. I knew I wanted to spend my life with Claire, but didn't feel ready for the other things that came with marriage and growing up. Where might we live? How would we live? How would I make a living? Claire was going through the same uncertainty. So far, I'd 'dealt with it' by just going climbing. I reasoned that I could tackle these big questions one step at a time and let a way forward emerge in due course. In reality, I was simply delaying grappling with these issues at my own cost further down the line.

Another thing I shared with Claire - although we expressed it in different ways - was a tendency towards low mood, lack of confidence and a negative way of thinking about ourselves, which had a long history running back to our younger years.

I was well conditioned from my time at secondary school to feel comfortable at the bottom of the school community, all sense of self-worth drained through humiliation, if there was much self-worth there in the first place. But it was when I started climbing that I became really conscious of this inner narrative of self-hate. Rocks and mountains offered me comfort and a means to carry on with life because of their indifference to me. I couldn't offend them. They could humble me, but never humiliate me. Mountains were somewhere I could just be, with nothing else required. The light they provided made it clear how dark my state of mind was in other environments.

It was on the way home from climbing that I could feel a pain returning to my mind. It was like the anxiety of travelling to an exam you know you are not prepared for and are certain to fail. You can't change the outcome, you can't hide from it anymore, and you're going to have to face the fact that you aren't up to standard. Like the practicalities of finding a place and way to live, I had a nagging sense that this problem would not go away on its own and I'd need to learn to carry it with me.

During this period, four things either pushed or pulled me from sinking further into a spiral of negative thoughts. First, as my connection to Claire grew from school girlfriend to long-term girlfriend to fiancée and I learned of her similar struggles, it was somehow easier to separate myself from self-destructive feelings. I had no clue how to adequately support someone who was going through depression, and I made many mistakes, but I could at least grasp that staying on top of my own would avoid making things worse for her. Second, on one of those trips home from climbing, I tried talking to a friend openly about just how dark I felt. Like me, he didn't know how to handle it. I didn't expect him to, but learning that others could not offer easy solutions to such a wicked problem did force me to develop a self-reliant way of working with depression, eventually even turning it to my advantage, at least on the cliff.

A third powerful antagonist to depression was simply climbing itself. I slowly came to realise that it was more than a source of acute enjoyment; it was also an immensely powerful therapy that could dissolve the darkest thoughts. Amazingly, almost everything about it had this effect. Shivering uncontrollably on a winter belay felt miserable at the time, but seemed to lift me for days afterwards. Hard training made my mind buzz with energy, even when muscles burned with fatigue. Laughs with climbing partners made me feel good. Solitude while working projects made me feel good. The warm sun, the cold wind, trees, rough rock, new places, sunrises, succeeding on routes, failing on routes; all of it lifted me, every day. And so I leant on it to prop me up. The backslide always resumed the moment I left these places, and so I kept time away from either a crag or Claire to a minimum.

It was inevitable that in particularly bad moments, my thoughts would lead down the obvious path of opting out of life altogether. All I'd need to do would be to let go. But my father had lost his father to suicide at 13. This was my final backstop against the dark trajectory I sometimes found myself starting down. Even at a young age, I could gather that the aftermath had affected my dad and his family for a lifetime. No doubt I didn't even know the half of it. The more I connected this tragic episode, far in the past, to difficulties my father still faced, it became an absolute red line that I would never cross. That thought process led to another inescapable conclusion: my life did matter to at least a few people, and as much as my brain wasn't wired to, I should take comfort and responsibility for the fact that they would be sorry to see me go. I could know this, understand that it was a fact, even if I couldn't *feel* it.

Although all these things kept dark thoughts in the background, they only treated the symptoms. They never reached down to the level of whatever was causing them to appear in the first place, which as yet I had no understanding of.

It was a joy for me to marry Claire. Both of us had a lot to hold our attention and deal with as we gained independence and began working. One evening at Dumbarton Rock, a couple of years after I'd started climbing, Peter and I had discussed what we might do for work when we left school. Peter said, 'You'll always just climb, Davey.' Naturally, I liked the sound of that, but I had absolutely no clue how to make it a reality.

There were a handful of professional climbers in the UK, but the most famous pros, Jerry Moffatt and Ben Moon, had spent years living on the dole. I'd done the same for two months when I left school, but the party was over on that front. It was a lot harder to live this way than it had been the previous decade. This was before the era of the social media influencer, and significant financial sponsorship still came after the world-class achievements, not before.

I was invited to join teams of athletes at Black Diamond Equipment and Five Ten. Getting all my ropes, climbing gear and rock shoes for free made a huge difference, but it still didn't pay for food and my bus fare to Fort William. I got a couple of breaks, though. First, the Five Ten deal did have a financial component. If I got a picture with the shoes or logo visible in the magazines, I'd earn £50 for a small image and up to £150 for a full page. It was £250 for a front cover, but that was rarely achievable because grit and limestone climbing was still far more fashionable than climbing in the mountains of Wales, the Lake District or Scotland. The advantage I had was that I'd started doing new routes almost continuously, and so most months I could bring in at least £50. Getting £50 after climbing a new boulder problem was a novelty that never seemed to wear off. Later, I got a bit of work writing for a website called *Scotland Online*. A keen Dundee climber called Alan Matheson managed the outdoor section of the site. He was a lovely man, very personable and persuasive. At the height of the dot com boom, he persuaded the company to spend some money upfront to make a dedicated Scottish climbing website that would become a go-to magazine and source of information. The site was fantastic, and it was the closest I've seen the Scottish climbing community brought together. The combination of Alan's energy and the investment to bring in writers was key. It was a huge shock when Alan died suddenly in an accident at home. I couldn't believe he was gone, and it made me realise how powerful a single figure can be in drawing together people who otherwise might not gel. Shortly afterwards, the dot com bubble burst and so did the website. But it gave me confidence that it may be possible to live as a climber, and it funded my first trip abroad, to Céüse in France, aged 21.

I bought a tent, a set of camping pots and a stove and excitedly showed them to Claire - my climbing partner on the trip. She seemed less excited than me. I just felt so utterly privileged to be able to go there and experience a climbing paradise I'd read about for years. I climbed until my hands were red raw and cut to pieces. The weather was terrible, with heavy snow and rain, and we were almost constantly cold, both at the crag and at night in our inadequate sleeping bags. I was so keen I didn't care, but it must have been grim for Claire. She held my rope for days while I threw myself at an

8b+ and learned a hard lesson that not all routes yield to finding a better sequence. I just didn't have the endurance. It must have been so obvious to her, but I only grasped it on the bus back to the airport.

After a few more trips to Céüse, I clipped the anchors on an 8c - my other lifetime goal. My whoop echoed around the limestone walls. I remember Niall McNair's reaction more strongly than actually climbing the route, when we caught up with him at the next sector along. 'WHAT?! You did it?!' I felt almost awkward to admit it. Niall is a better rock climber than me, with a lovely, intuitive style and athleticism. That I seemed to have impressed him was a strange position to be in.

I felt a little lost after this for some time. My two lifetime goals in climbing, E9 and 8c, both done. Beforehand, they had seemed like setting my sights on the moon. Then, just like that, they were behind me. I still loved climbing just the same, but it was like cycling over the crest of a giant hill on a bike and freewheeling down the other side, waiting for something to push against. This is a dangerous place for a young man with a tendency for depression. Without a project, my thoughts might slide anywhere.

Once again, something to push against came from someone else's vision. I'd started climbing with a burly Ayrshire climber called Scott Muir, who was keen to push his climbing but wasn't well suited to hard, fingery rock routes and didn't seem that keen on the dangerous stuff either. He always looked a little uncomfortable on steep rock, but put a pair of ice tools in his hands and it was as if he'd stepped into a different body. He was immensely strong and relaxed on steep, powerful ice and mixed terrain and was inspired by the 'sport mixed' routes in the valleys of Switzerland and Italy; sunny roadside crags with overhanging bolted dry rock leading to a curtain of beautiful ice. The routes looked amazing, and I was interested in them too. At the living end of the discipline, things got a bit weird, though. It was more and more about the steep rock and less about the ice. Some crags didn't even have ice and people were doing this 'dry tooling' discipline in the middle of summer. The approach taken by its protagonists, including Scott, was just to embrace the weirdness. Make it a feature, not a bug.

Scott went all in. At an old slate quarry at Birnam in Perthshire, a huge overhanging cave reached into the depths of the hillside. It already had one sport route across the roof: a completely manufactured 8b, climbed entirely on drilled pockets. It had been an abandoned project until I made the first ascent. The climbing was actually great; I enjoyed it so much, I went back the following week and soloed it. It was just so secure. If you could climb it, you could always climb it, so the rope wasn't necessary, right? I'd called it Hurlyburly, after Birnam Wood in Shakespeare's Macbeth: 'When the hurlyburly's done, When the battle's lost and won.' Only when the impossible happens

Left: Screaming Ab Dabs (E6 6b) on the Isle of Lewis with Scott Muir.
© Cubby Images

Right: Scott Muir in his favourite environment, continental-style sport mixed climbing with steep bolted rock and a hanging curtain of ice. Pink Panther (M9) at Ueschenen in Switzerland.

Above: Making the first ascent of Hurlyburly (8b) in Newtyle Quarry, Birnam. On the bus home to Glasgow, recounting the ascent in my mind, it struck me that the climbing was very secure and, even though close to my limit, could potentially be soloed if you could be sure of your fitness. I couldn't let go of the idea and soloed it the following week. Afterwards, a friend emailed me to ask, 'Did you really solo Hurlyburly?' 8b was the hardest I'd climbed in Scotland. I felt very fortunate that Claire had filmed my ascent. © Cubby Images

Left: An early repeat of Vertical Limit (M12) at Ueschenen in Switzerland with Scott Muir, at the time one of the hardest sport mixed routes in the world. Unlike Scott, I struggled to relax while climbing on tools and lacked his body strength. But I could make up for this to a degree with experience of grinding onwards on Scottish winter leads. I took over an hour to climb across the 45-metre roof, eventually fumbling a move on the three-metre finishing headwall and dropping an ice tool. Still hanging on with the other, I improvised a few hand jam moves in a flake and mantled my remaining ice tool to clip the belay. © Scott Muir

and Birnam Wood comes to Dunsinane does everything fall apart for Macbeth. Absolute certainty in anything might not be a good idea.

The deeper half of the cave got much less light and wind and was nearly always soaking wet. The rock was no use to anyone, so Scott bolted it and drilled smaller holds than those on Hurlyburly, this time for ice tools. I belayed him as he worked on it. Trying the moves for myself was a sharp contrast to watching Scott - I struggled to feel comfortable, disconnected from the rock on the end of my ice tools.

It was yet another completely different sport for me to get involved in, and between sessions I stayed at Scott's house just down the road from the quarry. Scott was sponsored by Red Bull, an association that was a red rag to the traditional Scottish climbing scene, making him a polarising figure. Behind his front door he had a floor-to-ceiling stack of 24-can slabs of Red Bull. When the alarm went off each morning, he'd spring bolt upright in bed, grab a can from the bedside table and neck the whole thing. With one hand, he'd crush the can and let out a loud, deep 'AAAAAAGGGGHH!'. With the other hand, he'd hit play on the stereo, which would belch out house music, and seconds later he'd start jogging around the house, preparing kit for the day's climbing.

Scott had a lot of energy and felt he had a contribution to make to traditional Scottish winter climbing, but he just wasn't suited to the painfully slow onsight style, inching up rime-covered E1s, fighting to dig for a runner worth the name. He kept ranting about 'fucking turfy corners' and how he wanted to do a much harder route, but in redpoint (rehearsed) style, working a climb on the top rope first to find the hooks and gear. It would be a major departure from the incumbent Scottish winter climbing ethic of ground-up, onsight climbing. To justify the break from tradition, it would have to be a huge leap in difficulty. Something hard enough that no one could say, 'Someone could have onsighted that.' It needed to be outrageous.

He'd found it. On one side of Fiacaill Ridge in the Northern Corries of the Cairngorms was a plumb vertical 35-metre wall with a bold summer E4 running straight up the centre. The route was called The Hurting, and it had a reputation for being pretty tough and scary, even in summer. I thought it was a legit crazy idea. It was so far above what anyone had attempted in winter before, and it struck me as ridiculous. But I was curious, so I agreed to belay, slightly sceptical that this project was realistic for him. I just wanted to witness someone on that sort of terrain in winter.

We stomped in on a standard blustery Cairngorm morning, and Scott hung a rope down the line and worked the moves. He could actually do most of them, and it was incredible to see him climb through what seemed like a bald, impenetrable wall, plastered in rime and verglas.

Back at his house that evening, Scott was practically bouncing off the walls with nervous energy. He felt that a lead was on - an outrageous prospect. I was impressed that he'd got himself to a point where he was actually considering it; it made me nervous even to think about, despite the fact it wouldn't be me on the sharp end. My hunch was that he hadn't done enough leading routes of Grade VIII or above to have the broad foundation and consistency in his climbing needed to make such a big leap in difficulty. I recognised this feeling myself. But I had also now seen enough professional climbers in action to know that it is short-sighted to write them off just because things seem stacked against them. If the fire is strong enough, it can sometimes burn through.

Observing Scott as he prepared for The Hurting, I noted that he was clearly intimidated as well as excited, which mirrored my progression through the grades in winter climbing, and I was fascinated to see how he might overcome this. The keystone of my own confidence in winter had come from a piece of good fortune that I'd only recognised with hindsight. When I started university, I'd climbed a couple of Grade VIs but was still consolidating at Grade IV and V. I'd started climbing a lot with Andy McIntyre, also from Ayrshire, who was really good on ice and mixed, but like me occasionally got stressed at the prospect of getting out of his depth on winter climbs. My 1,000-footer on Ben Nevis was still quite fresh in my memory, so I shared his willingness to back off if things seemed likely to get out of hand. But winter climbing is so different from rock climbing, where if you grind to a halt, it usually happens quite fast, in seconds or minutes. You can't just hang there as long as you like. The clock is ticking and either you engage reverse gear quickly, fall or jump off. In winter, when you grind to a halt, you can often just keep grinding, not getting any higher, but still probing, gradually admitting defeat at the same time as trying 'one last thing' over and over. Crucially, I found that I could tolerate that uncertainty for a bit longer than Andy.

Often, the discomfort is not just felt when you are out in the lead, but after leading a pitch. While forcibly tied to a belay for a long time, you can only stare upwards, intimidated by the terrain above and its refusal to give up its secrets until your partner finally overtakes and inches ahead on their lead. The impulse to retreat is normal, but also hard to act upon. No one wants to be the half of the partnership to admit defeat and ask the other to go down. A halfway house is to ask your partner if they wish to lead again, putting the ball in their court. If they feel as intimidated as you, a mutual decision to retreat is easier to swallow. If they opt to lead, that's up to them. I didn't mind retreating, but not without trying first. The more I opted to lead, the more Andy asked, and the more I expected to lead everything. It was the best climbing intervention I could have asked for. I grew to associate all climbing with leading: if I wasn't leading, we

Above: Andy McIntyre on Ben Nevis.

Right: Crest Route (VI,6) on Stob Coire an Lochain, Glen Coe with Andy McIntyre.
© Cubby Images

Prore (VIII,8) in Coire an Lochain, Cairn Gorm.
© Cubby Images

On the second ascent of Viva Glas Vegas (VIII,7) on The Cobbler with Andy McIntyre.
© Andy McIntyre

Top right: Climbing Riders on the Storm (VI,5) on Ben Nevis with Andy McIntyre. Ben Nevis is famous for its thinly-iced slabs. © Cubby Images

Bottom right: Topping out on Riders on the Storm. © Cubby Images

weren't going up. I started to hate the impatient wait on a belay and grew to embrace the uncertainty of the terrain above and engage fully with the process of unravelling it, piece by piece. I ground to plenty of halts. But as I became the default leader in my partnership at the time, *I'd* have to own the decision to announce retreat, and so I tended more and more to just keep grinding, eventually realising that you could grind out of halts as well as into them. After two winter seasons, I'd got to Grade VIII and was acutely aware that I'd never have done so without that sheer volume of time tied to the sharp end of the rope.

Scott Muir worked as a winter climbing instructor and had more leading experience than me. But hard climbing is never so simple. I had already noted among other mountain guides and instructors I climbed with that (with some exceptions), if anything, their experience could make them more uncomfortable with operating at the very edge of their abilities. When it comes to cultivating an appetite for fear and intimidation, experience is a double-edged sword. If accumulated on terrain well inside your comfort zone, that comfort zone may actively contract. To be effective, experience must include volume of psychological discomfort, not just volume of climbing. As we walked in to The Hurting in the early morning gloom, I admired Scott's resolve to swallow such a large dose of fear in one go.

We both abseiled down the line from Fiacaill Ridge. On the way down I could see the odd hook or gear placement that Scott had scraped out with his ice tool. It still looked ridiculous, like a different sport from the big cracks and turfy corners I was used to. Scott top-roped again, and that same curiosity wormed its way into my thoughts as I watched him. *Could I do the move that way? What about that little seam just a bit to the right?* I became uncomfortable watching him, unused to being a spectator of winter climbing and anxious to discover these secrets for myself.

Scott asked to come down, and he was unusually silent for a few moments as he donned his belay jacket. Then he surprised me by saying he realised that leading the route was not going to happen for him - it was simply too dangerous. I was impressed that he could make such a clear decision and could sense that he immediately let go of a lot of tension. He then asked if I wanted a go. Last time, I thought there was no point. Now, the curiosity was eating me, and I desperately wanted try a few moves. But I wasn't ready to top rope a winter climb. It just wasn't what winter climbing was for me. I wasn't against rehearsing moves in any climbing discipline, but winter was utterly synonymous with leading, given my personal history with it. Through the blizzard I'd only really been able to see the details of the moves in the first few metres. I asked Scott if I could tie into the lead ropes and try the first ten feet or so over a bulge. Even that bit as a little boulder problem looked brilliant, if desperate. I could feel the first few hooks and then climb down. I just wanted a taste of what I'd witnessed Scott climb.

I clambered onto the block we were anchored to and reached up, scraping the edge of my pick in the back of a rounded depression. I yanked hard onto a straight arm, willing it to rip. As thin as the placement was, it wouldn't rip, and after a little internal discussion had run its course, there was no choice but to pull up on it. Then, a ridiculous sidepull hook in the back of a thin groove, standing on nothing but rough granite crystals with the front points of my crampons. Then another crazy move, turning my ice axe upside down and jamming it against a rime-covered undercut hold, leaning out and building my feet up. Scott had told me there was a tiny incut bump somewhere on a rounded break above. If I could find this placement, the move through was secure enough. But at full stretch, it was impossible to tell if my pick had found the bump. I blindly reached and scraped. Is that it? I yanked on the tool. It wouldn't rip. What was I going to do? I hadn't ground to a halt yet, so the hardwired criteria for retreat were not met. I had to pull up.

It had taken 20 minutes to fight my way just a few body lengths above the snowy ledge at the base of the wall. But this had already been more intense climbing than I'd ever done with ice tools. I was vaguely aware of Scott shouting encouragement from below through nervous body language, standing off to the left in case my 26 sharp steel points cratered into the ledge. The wind burned my face, driving sharpened snow crystals into my reddened cheeks; a harsh reminder that I was losing focus to thoughts of whether or not I should be here. The following hour is completely lost to my memory.

The next thought that entered the record was that I'd ground to a halt. A fat lump of frozen turf beckoned one metre out of reach to my right, and above it lay a few metres of easy climbing to the top. *How the hell did I get here?* That wasn't important. One more hook and I'd reach that turf. From a horizontal torque, I built my feet up to use the full reach of my tools. I focused my gaze hard on my feet to avoid looking at the ropes hanging free to an optimistic micro-cam ten feet below and the first actual protection much further down. I knew I was gearing up to take a monster fall. How else was this going to end? As my arms wilted, I tried everything to move my body rightwards. One option remained: a rounded depression I knew would rip. I removed the right tool, pulled on the left, and it ripped.

Crystals of snow battered off my Gore-Tex jacket as I accelerated down the wall. *Just wait another split second; things are about to get a lot faster.* But full speed never came. With a zipping sound from one of the skinny half ropes, I slowed, stopped and

Opposite: Moving above the initial thin groove of The Hurting (XI,11) in Coire an t-Sneachda, Cairn Gorm.
© Steven Gordon

boinged upwards again. Scott made some sort of whooping noise. I looked up and blurted out, 'That fucking cam held!' After some adrenalised laughter, I hauled up the rope to the runner and stopped in my tracks when I saw that it was holding me on only two of the four cams, having slid to the very edge of its icy slot. I stopped laughing and asked to be lowered very slowly back down to the better gear below.

I had to finish it now. The following week I returned and fought my way back to the same spot, staring at the hairy lump of turf, one metre out of reach up and right. I was acutely aware that whatever I did, it would have to be different from all the options I'd tried last time. I attempted to relax and consolidate, to notice any secret rock features I'd missed last time. I reached left to try to get a better temporary resting position on my feet and found a slightly more positive hook. As I rested, I tickled my tool up and left again. Perhaps there would be a useful runner up there? There was no runner, but there was another hook. Instinctively I pulled on that one, too. It was the wrong direction. Now I was looking directly across at that turf, almost three metres to my right: my arm span plus my ice tool span. Torquing the end of my pick in the hook, I lurched rightwards as far as I could and, at the very first lunge, whacked my tool into the turf. Done.

Except that I had no more span to un-torque the left tool. I felt ridiculous. Literally hanging onto the finishing ledge, spanned out in an Iron Cross position with my face pressed against the crusty rime coating the wall, I was still going to fail on this climb. For a few long minutes, the only option seemed to be to leave behind the left tool and attempt the final metres using the other. But shuffling my weight between each arm to try to ease the pump suddenly gave me the idea to drop onto my first two fingers on the right tool, my first two fingers on the left, and that gave me a couple of centimetres to release the torque and scurry to the summit.

At the top, the forecasted 100mph winds demanded a sense of urgency, and provided a convenient reason to delay thinking about the outcome of my lead. Fighting to stay upright in the gusts, I prized apart my frozen eyelashes, arranged a hurried anchor and descended to make a quick escape into the coire. As we stomped through the whiteout, isolated in our own thoughts by the roar of the wind outside our hoods and goggles, I could no longer avoid thinking about the success of the ascent and that I'd probably have to apply a new grade to a route so hard. I felt a sense of dread about this, which was more intimidating than the route itself. Not only did I not have Scott's talent for climbing with ice tools, he had provided the vision, a lot of the logistical problem-solving and handed me the lead with the back of the route already broken. Moreover, on routes of such a high grade, I was used to a lot more failure, and

Right: After completing The Hurting. Looking forward to putting my hood up and hiding from both the blizzard and the prospect of having to apply a new grade in Scottish winter climbing.
© Steven Gordon

Opposite: Good protection and the last security before launching up the headwall of The Hurting.
© Cubby Images

the fact I'd taken only one fall left a sense that I had not really earned a route I knew would be seen as ground-breaking by some others.

A few days later, Scott and I returned with Cubby, who would take pictures as I climbed the safest section of the route again. As we rigged anchors and uncoiled ropes at the top, a mountain guide walked past with clients, obviously curious about what we were up to, rigging ropes on a wall away from the established climbs.

'What route's that you're on?' he asked, slightly dismissively.

'It's The Hurting,' Scott replied sharply, reflecting the dismissive manner right back.

'Never heard of it. What grade is it?'

Scott gave a one-word answer: 'Eleven.'

The guide gave a look as if Scott hadn't understood his question and wandered off into the mist. I wanted to bury myself in the snow and hide. We knew before we even started that The Hurting was significantly harder than the handful of Grade X routes that had been done so far, but Scott saying it out loud with such confidence still made me squirm.

The climber who had done two of those Grade X routes was another Ayrshire-born man called Alan Mullin. Over the past couple of years, I'd been reading about his exploits in magazines, and they almost made me feel like giving up winter climbing. He described epics and sticky situations that made my stomach turn. If that was where

the sport was going, I wasn't sure I wanted to follow. I wasn't alone either. I'd heard of several good winter climbers who'd quietly sworn they'd never climb with him again. Eventually I had little choice. Alan rang me out of the blue and informed me that we were going to Lochnagar the next day. Having heard so much about him, it was odd to hear his voice on the phone for the first time. He sounded so young for someone who'd had such incredible epics. 'We'll do Mort, mate; it'll be fuckin' raaad.' He reminded me of kids at school, his hard-edged working class Scots accent grating and yet very familiar.

He picked me up in Inverness, and we drove across the Cairngorms under a very cold starry sky to the car park at Loch Muick above Balmoral Castle. The journey was exhausting. Alan practically bounced out of his seat with excitement, spit splattering the steering wheel as he recounted extraordinary failures on routes either with partners or on his own. 'I'm not climbing with FUCKWITS anymore!' he spat, and then tipped his head back and laughed crazily. I'd been asking him why he'd been soloing so much. Alan had been using a system of leading with the rope while climbing solo. Various systems and devices can be used in place of a partner, but all are fickle and prone to problems, and nobody was doing the hardest routes in this style. Moreover,

Right: Alan Mullin waiting for winter.
© Cubby Images

the system Alan was using was one of the worst, leaving the auto-locking belay device, a Gri-Gri, at the anchor, out of reach from him if it jammed, which it often did. He explained that he'd had one falling out after another with many of the best winter climbers in Scotland, often when they refused to let Alan carry on when things were getting out of hand. That explained his phone call out of the blue.

Alan candidly explained why he was on a mission to prove his skills as a winter climber, both to himself and others. School in deprived Ayrshire had been chaos and had ended early after he broke a chair over a teacher's head. He believed that joining the Army had saved his life. I'd already noticed his neatly prepared climbing gear and clean, folded jackets when I'd loaded my rucksack into the boot. The Army had been everything to him, and when they rejected him after atrocious handling of a severe back injury he had sustained during training, he was devastated. He wanted to show that he could go as far, if not further, than anyone in the winter mountains. I wasn't sure if he'd been diagnosed with bipolar disorder by then, and he didn't tell me.

However, winter mountains are not a straightforward adversary. Even if you are looking for a fight with them, they may not offer one. We arrived at the car park at 11pm, and I walked a few metres in the frigid air to go for a pee and watched a big shooting star cross the sky. The silence of the glen under a deep winter frost was a welcome contrast to Alan's three-hour rant. By the time I'd walked back to the car, Alan was in his sleeping bag, having fired up a joint. He launched right back into stories of fights with climbs and people. By midnight, he admitted that we should maybe sleep and set the alarm for 2am. There were ten seconds of silence before he sprang upright and shouted, 'FUCK IT, will we just walk in NOW, mate?'

'No, Alan, I'd like to sleep,' I said gently but firmly. I was surprised when he accepted my response.

At 5am we stood below Mort as it revealed itself in the orange dawn. The critical rule of Scottish winter climbing is that the climb must look wintry, i.e. white, to be considered in acceptable condition for an ascent. The steeper and harder the routes get, the harder it is to find them in condition. 'Fucking BLACK, again!' Alan's rant echoed round the coire. I was too sleepy to care. As he charged up a Grade VII nearby that did look white enough, I fell completely asleep belaying and woke to him yanking the rope and screaming at me for slack. *Another fuckwit*, I thought. Alan seemed too angry at our chosen line being out of condition to be annoyed with me. On the walk out, he ranted continuously as I jogged along behind and nodded.

Shortly afterwards, Alan rang again and announced we were going to Glen Coe to climb The Duel, a Grade IX first climbed by Cubby. This time when we met, Alan

Left: Alan Mullin setting off tentatively on pitch two of The Duel (IX,9) in Glen Coe. He soon hit his top gear and raced up the rest of the pitch in 20 minutes, placing only two runners.

seemed to be operating at a fifth of his usual pace. He was quiet, hunched over and seemed unsure about whether it was worth walking in. We did, however, and as he dawdled ahead of me on the path in the dark, I almost lost an eye about 20 times, bumping into the ice tools strapped to the back of his rucksack. This was a different guy from last time. He wasn't feeling it. The mood grew still more indifferent in the coire as it became clear that the climb was in seriously borderline condition. It definitely wasn't wintry looking from a distance, but we walked up to the foot of it anyway, and it turned out that the wall was coated in a layer of clear verglas, so we uncoiled the ropes. It would have been better not to climb it. It was one of those 'you had to be there' days, and I should have appreciated that others would say it was not in condition. Alan announced that he didn't care what anyone else thought, even though he spent a lot of time talking about what they were saying about him. I climbed the crux first pitch, which was absorbing and tricky to protect given the verglassed cracks. Cams are useless in verglas conditions, and if the temperature is low enough, even nuts cannot be forced to seat in the cracks with the pick of an ice tool.

When Alan followed, he seemed caught between two extremes of mood. He continued to rant about the fickleness of Scottish winter climbing and the difficulty of dealing with 'fuckwit' partners. At the same time, he seemed to be enjoying the climbing itself and was moving well. The next pitch still looked hard, and it began to snow heavily. Given Alan's unstable mood, I contemplated the worst of all winter leading scenarios. I wouldn't mind if Alan wanted me to lead again, despite still feeling tired from a long, hard pitch below. But since Alan was a much better climber than anyone I'd ever partnered with in winter, I worried that he might attempt the pitch and get high on it, then decide to retreat and pass me the lead. I really felt the cold on hanging belays, and although I'd likely just get on with it, I'd rather lead now if I was going to at all.

Alan still sounded low but confirmed he would lead through. What followed astounded me. He scraped awkwardly up the first move or two of an offwidth crack. His foot slipped, and he apologised for sketching around right above my head. I sensed that he was self-conscious and not in the right frame of mind for a lead. But suddenly, he hit a gear that I'd never seen before or since in any climber. He started to flow up the crack at lightning speed. Everything about him flipped. Even his breathing was different. I watched open-mouthed at the speed and confidence with which he swarmed upwards, placing two runners in 40 metres. In 20 minutes flat, I heard him shout 'Safe, mate' from the top of the buttress.

Walking out, I complimented Alan on his performance and asked what made

Opposite: Alan Mullin seconding the crux pitch of Happy Tyroleans (VIII,9) in Coire an Lochain, Cairn Gorm. Alan's car needed a new starter motor so he had to park it facing downhill at the ski car park. After we finished the route and got the car slowly rolling down the hill and jumped in, a herd of reindeer strolled across the road, blocking our path. Alan told me to steer and leapt out of the car, pushing reindeer out of the way as the car slowly picked up speed.
© Cubby Images

him climb so fast and not even bother to place protection. He explained, in his own particular style, that he'd been feeling so angry and fed up with winter climbing that he just let go and went for broke. I guessed that many of his well-known solo successes and failures in winter were likely borne from the same mindset. It was scary to watch. Alan always seemed to be on a knife edge, but I couldn't help notice that he seemed to find a way to convert deep frustration into impressive success at least some of the time. I wondered if classic sport psychology techniques of reinforcing self-belief and positive thinking could ever work for someone in Alan's position. It struck me that this approach could go badly wrong for him.

Back in Glasgow, Alan and I took some criticism for climbing The Duel that day. I explained to our critics that the wall was verglassed, and although it didn't look white, it still fell within the rules, even if it was on the edge of them. I think their minds were made up already, probably fairly. Alan rang me. 'Welcome to my world, mate!' he screamed down the phone before laughing crazily. Like Alan, the episode did destroy part of my motivation for winter climbing, and ever since then I have focused more on bouldering during the winter. It felt just a bit too close to the pain of mainstream sports, where my effort had to be judged through the lens of someone else and the bias they carried. I thought I'd left that behind when I started climbing. While Scott was trying The Hurting, I witnessed the sneering he received for deviating from the style of onsight climbing and practising moves on a winter route. Onsighting is indeed a highly valuable tradition and well worth protecting and celebrating. Where I differed from Scott's critics was that they seemed to feel that any diversification of style in winter climbing had to be stamped out and that practising moves inherently diminished the experience.

I wrote an article for one of the climbing magazines to try to present my objection to this notion and defend the benefits of a range of climbing styles. For one thing, the main criticism of pre-practising hard climbs, so-called 'redpointing', was often borne of sheer ignorance of its challenges. It was often said that pre-practising a climb removed all unknowns and made the ascent a foregone conclusion. In fact, it was the opposite. Climbers would use this method to target climbs so hard that a feeling of uncertainty was almost inevitable. If you try the moves on a climb but cannot do them, then clearly success is not guaranteed. This point is the start of a process of exploring those unknowns, otherwise known as an adventure. If you climb the route onsight, you don't really know if it's genuinely hard enough for you. Often, you have to err on the side of easier climbs for safety's sake. If it's too easy, there isn't any doubt. If it's too hard, the doubt lasts only as long as it takes to get pumped. You fall off and that's that.

The onsight is gone forever. With redpointing, the challenge of the climb remains even after you fail on any attempt. It will always be there waiting for you, perhaps hanging over you, if you are willing to accept that you are truly committed to finishing it. If you decide it is too hard and walk away, there is the question of whether you really wanted uncertainty in the first place. Trying the moves on the safety of the top rope during the preparation stage makes uncertainty a given because it allows climbers to lift their sights to climbs so hard they might barely get off the ground on the initial attempts.

I didn't top rope The Hurting; I only watched Scott do so. But I would never have even tried it had it not been for Scott's approach to the climb. I realised that some of the sneering at characters such as Scott and Alan was fuelled by attitudes to them as people - both difficult in their own ways - rather than what they did. It's fine not to like someone, but I was uncomfortable seeing them treated with disdain or even contempt because of their personalities and chosen climbing styles.

One of their critics told me he was seething after reading my magazine article. I tried to expand on my thesis that climbers using all these different approaches were looking for an adventure and urged caution about assuming that it was about removing difficulty. I just felt that people like their difficulty served in different ways. Scott, Alan and I were all trying to find our limits on routes, in ways that matched our abilities, attributes and limitations. I learned so much from both of them and I was, and still am, anxious that the activity of climbing always remains wide enough to accommodate colourful characters who do things differently.

I climbed with Alan only a handful more times. His mood swings seemed to be getting more severe, and the next time I heard from him, it was clear things had become far worse. Alan was consumed by mental illness, with severe bipolar symptoms and psychosis. After a calamitous episode with health and social services, he took his own life while in custody after a psychotic episode.

Mental illness is all around us, causing pain, suffering and death. I'm not sure why it still surprises us how serious it can be when it takes friends and family from us. Difficult as it is, I resolved to respond to the loss of Alan and others by constantly trying to improve my understanding of the factors that influence it. At that time, especially right after Alan's loss, I felt quite helpless. Moreover, I could not avoid the observation that some of the narratives Alan had expressed as we walked through dark, snowy mountains also rumbled in the back of my own mind. These were feelings of insecurity, worthlessness and a constant background hum of self-destructiveness that made no sense and had no obvious source. However, I could see differences between Alan's and my own ways of coping that seemed to be working for me.

Right: Alan Mullin lamenting another day of poor conditions, "It's fucking black, again."

The combination of the presence of family and friends, education and the opportunities that spring from it, as well as practising a broad spread of climbing disciplines, all seemed to help me stay afloat and maintain my ability to find satisfaction and joy in life. I wasn't measuring my worth through what came out of someone else's mouth, no longer placing my eggs in this basket. When winter climbing was frustrating, I went sport climbing or bouldering, or just trained. I could get a change of scenery and spend time with different people if something was getting me down. All this made it easier to maintain an even perspective and to absorb knocks, but critical as it was in helping me move forward with my life, it was still only treating symptoms. Getting to grips with the causes of depression would come much later.

I also learned from Alan that, despite the popular narrative, a negative state of mind could be an effective, if dangerous, performance tool in the right circumstances. The mark of an expert or 'professional' athlete is the ability to produce consistent performance at a high level in the face of the rigours of life and all its setbacks. Traditional sport psychology suggests, with weak evidence, that positive thinking builds a foundation strong enough to withstand and power through these setbacks. Climbing with Alan gave me a hint that this paradigm may be over-simplistic and that endlessly chasing a mindset I could seemingly never achieve may not be a prerequisite for my best performance. The challenge, however, was to find a way to make my own temperament work more consistently in my favour than it had done for Alan.

DUMBY DAVE

Previous: The Requiem face from Dumbarton High Street.

Having climbed more routes at Dumbarton Rock than anyone else, other locals sometimes called me 'Dumby Dave'. I spent a lot of time explaining that I was actually Glaswegian. As Claire and I started to work out where we could afford to live, we joked that it would be funny if we got a house in Dumbarton. Pretty quickly, that became not such a bad idea. Claire had a small salary from her job managing an optician's shop; I was just doing the odd day of work here and there for climbing magazines or setting routes. Our options in Glasgow were limited, but flats were cheap in Dumbarton since for the average person it wasn't the most desirable place to live. It was already a second home to me, and I had no objection to making it actually home.

We chose a tenement flat, put in an offer and looked into mortgages. I felt completely ignorant of how debt worked, so I started by searching for what the word mortgage means: 'Contract until death'. Sounds great. At least you know where you stand, working in service of the bank for a lifetime. I looked at how the interest on a loan is the part that is serviced first, with the capital diminished only in the later stages, and at the vast sum of interest I'd pay over the life of the debt. It was eye-watering. I would like to say that I was educating myself to make sound choices as I entered the world of borrowing, but in reality, fear drove my research. Perhaps I could find a reason to keep hiding from this headache. I was dismayed and angry that I was entering a rigged game, and swore that if I ever earned enough, I'd plough all of it into the mortgage as early as possible and avoid playing by the bank's rules. The fear did at least serve its purpose: I learned that there was an alternative way to play the game, and this, in the end, gave me confidence to take the next step.

The upside of choosing Dumbarton was that our mortgage payments were only £80 each per month. I ought to be able to continue as I had been, climbing lots and working only when the weather was bad. Claire was at work the day the sale went through. I picked up the keys and dropped off a few things at the flat, then headed straight back out for a session at the boulders before darkness fell. When Claire finished work, she wasn't best pleased. She felt I should have celebrated our milestone by spending a few hours at the house, something she laments to this day. I had felt that celebrating was what I was doing. I had bought a whole way of life, one that involved walking the ten minutes back and forth between the boulders and home. Today, I'd do it differently.

One of the things I liked about the town, something that made me feel at home before I even moved there, was that it didn't take itself seriously. Those who wanted to move out of Glasgow and had a bit of money chose the much wealthier Helensburgh, another 15 miles down the Clyde. It seemed that few people genuinely chose

Dumbarton; Dumbarton chose them. So nobody gave a shit. The down-to-earth attitude was refreshing, mostly. People here are delightfully friendly, with zero hang-ups about status.

I immediately settled into my daily commute between the house and the boulders. I hated climbing early in the morning and had usually stirred enough to get out the door by noon. First, I'd walk past Dumbarton's High Street shops, which were closing one by one and being replaced with payday loan shops. They were just a smidgin less predatory than the scary men on street corners they competed against. Beyond the end of the High Street, I'd pass the side entrance of the sheriff court. Defendants or their relatives would often be stood outside this doorway, smoking and waiting for proceedings to resume inside. 'Want any eccies?' a teenage lad once asked me as I walked past.

'No thanks, big yin,' I replied. 'All the best, mate,' I continued, under my breath.

I'd often see an old woman leaving Morrisons supermarket and whirring up the road on her mobility scooter. It had surely been tampered with to remove the speed limiter - she'd be shifting. She'd always have finished her first fag by the time she passed the Denny Tank Museum and, with one hand on the handlebar, she would flick it into the same bush and press on, pedal to the metal.

Climbing through a hole in the fence behind Morrisons, I'd leave the noise behind and stroll across the tangled metal of the still derelict docks to arrive at the boulders. Often, I wasn't the only one escaping life in the town. On the first boulder was a big cave, and I'd been exploring the idea of a climb going across it for some time. The rock was soot black from endless driftwood fires, and there was broken glass and frazzled beer cans everywhere. I'd started taking a bin bag with me on every session, and after I'd finished climbing it didn't take long to fill it and dump it in the public bin by the road. Others had seen me doing this and gradually joined in. Another climber called Rab once tugged at a black plastic bag just under the surface of the mud and gravel beneath a boulder. It wouldn't pull, so he dug it out to find that it contained six identical hatchet axes, presumably in storage for later use. No one ever forgets where they buried the hatchet.

One day I was playing on the moves in the black cave when a man in his 50s appeared and asked if I minded him joining me there. It was his regular spot, just as it was mine. I said, 'Of course not,' and he sat, placing beside him his white carrier bag containing ten cans of Tennent's Special Brew, a very cheap, strong beer specifically formulated for escaping your life. In the time it took me to tire my fingers out in the roof - about two hours - he'd sunk all ten cans. If I hadn't seen him do it, I wouldn't have

Right: Rab with hatchets.

believed he'd drunk a drop. His speech didn't even slur. As I tried moves and he drank, he told me his life story, which was heart-rending - a tale of poverty, loss, prison and loneliness. Yet he also seemed content and cheery, even before the Special Brew had kicked in, if it did at all.

The young folk of Dumbarton also escaped to the boulders. I once arrived for a training session to find a group of lads had set up base camp in the black cave. Their fire wasn't in the cave though, since their fuel - huge timbers and logs that had washed up in a recent storm - was too heavy to carry up the beach. That wasn't all. They whooped with delight when they found a large petrol drum that had also washed up and was apparently full. Rolling it nervously onto the fire, they danced about from several metres away until it blew up in a bright fireball and flattened them all.

Climbers visiting the Rock from further afield sometimes found the hard-edged local young men intimidating, but I never did. Without climbing to entertain them, they tended to rely on some mix of alcohol, mischief and banter with whoever was around at the boulders. It's true that these youths continuously assessed those around them, placing people in one of two categories: those they could potentially beat up, and those who could beat them up. Banter was a key assessment tool. 'Aye, just 'cos yer built like a tank, dain't hink a wilny take ye,' one reminded me - less a threat and more a recognition that we could respectfully coexist among the rocks. Although climbers were often small, skinny and softly spoken, they went up cliffs and therefore might be mental. It was not a safe bet to place them too confidently in either category.

The boulders were covered in a mixture of new and very old graffiti. The older stuff had been crafted with some care and was quite artistic: the eagle emblem on the Eagle boulder and some cryptic political ramblings across the overhanging wall of Pongo. The newer stuff was just messy, and it occasionally ruined the friction on the holds of the classic climbs. I appealed to one group of kids with spray cans to avoid painting the boulders and stick to the decaying man-made walls of the old sawmill instead. 'How no', there's hunners?' asked one. I'd expected a sharper response, but he seemed a little on the back foot since I'd asked him nicely, as if he wasn't used to being given agency. I could relate.

On rest days I sat on the sill of my broad window bay in the flat, indulging in endless daydreams about moves on projects while waiting for the rain to stop so I could nip round to the boulders. It was also a good place to watch the vibrancy of Dumbarton life in the street below. One evening a young man turned into the street, walking at pace down the centre of the road, carrying a large iron bar. He stopped abruptly in front of a Rover parked next to Claire's car and stood for a moment, staring at our tenement block with some sort of intent I couldn't read. Then he turned, raised the bar over his head and smashed every window, light and panel of the Rover before retracing his steps up the centre of the street. After a good 20 minutes had passed, I gingerly crept out to check if Claire's car had been damaged. Her brother had moved abroad and sold it to her at a nominal price, otherwise she'd never have been driving an Audi, and she loved having it. A man was hovering in the next close along from mine, and I went over and asked if the Rover was his. 'No,' he said, shaking like a leaf, 'I've got one exactly the same as that, but I park it two streets away since I know they're after me.' I didn't ask any more.

As the rain cleared one morning, I was making a cup of tea to start the long process of waking up before my commute when the kitchen lights started buzzing and flickering. A short time later, aware of noise outside, I retreated to my window perch and saw the street blocked with fire engines, ambulances and police cars. Paramedics were tending to an injured man, blackened by what looked like soot, sitting on the foot-high sandstone wall outside the opposite tenement block. Half his shirt was missing - I presumed the medics had cut it off, although it looked almost like it had been burned off - and the skin of his exposed arm was red and scraped along its whole length. Police were taking statements and men who looked like building contractors stood

Finishing up the easy slab of BNI in May 2003, during the first ascent of Sabotage (Font 8A), which takes the right side of the roof below.
© Cubby Images

around nervously. After the injured man was taken away in the ambulance and the other emergency services left, the builders lingered. One of them appeared from the close, holding a section of thick cable. Looking nervously up and down the street to see if the coast was clear, he flung it into his truck and they all drove off. I guess you have to be careful what you cut when stealing the copper cabling supplying the whole street.

Even in winter, with the rain lashing daily, I could see the Requiem face from my windowsill seat and I watched for any opportunity to climb. If the rock was wet, it would look deep grey and unappealing. But at some point most days, the weather front would pass over and the wall would light up orange in the late afternoon sun. I'd run down the steps to my climbing gear cupboard in the stairwell, grab my stuff and walk round. The cupboard on the landing would have been a toilet in the days before they were plumbed into the flats themselves, and mine was rammed full of all the climbing kit I owned.

Returning from a climbing trip, I couldn't find the key. Claire had left it in the door while in a panic to get our ladder after our upstairs neighbour flooded our flat for the third time that year. There was no key in the door now. I picked the lock to find my cupboard was serving as accommodation for some poor soul who'd obviously been searching tenement closes to shelter in and had got a lucky break with mine, with a metre square secure space to crouch in. A pair of manky socks had been hung up to dry, and there were empty tins of cheap baked beans and creamed rice lined up on a shelf. It was a grim thought that it could be a sanctuary. Some kit was missing: ice screws and a brand-new pair of boots. Scott Muir and I had been planning to compete in one of the ice climbing world cup competitions in Switzerland, and my sponsor at the time had given me a very loudly decorated and specifically designed pair of comp boots, shiny silver with completely rigid soles. For weeks I eyed strangers in town, expecting to see someone walking down the road in the shiniest footwear ever to grace Dumbarton's High Street. Wearing those boots, he may also have needed the ice screws for self-defence.

Much of my time at the Rock was now spent trying the many obvious unclimbed lines on the boulders, and a few on the main face. I was slowly learning not to underestimate how much I could leverage my fascination for working out the details of moves. Malcolm Smith had started climbing there a lot during the winters, and he completed a line I was also trying, which he called Supersize Me. The name appealed to my ongoing struggles to remain lean, and it was the first climb at the Rock, and indeed Scotland, to get the big grade of Font 8B/V13. I couldn't touch Malcolm's sequence

Top right: Leaving my tenement flat in Meadowbank Street, Dumbarton to go to the boulders.

Bottom right: My windowsill seat view of Dumbarton, with Inverleven Distillery and the Requiem headwall catching the evening light.

Opposite: Working the final moves of the Requiem headwall in May 2005.
© Steven Gordon

for the crux, but after working out a series of bizarre heel hooks and counterintuitive hand movements, I started to get close.

Supersize Me is 30 moves long, and right at the end you get a brief rest on a good flat hold at the start of Slap Happy, the exit problem on the face. One busy day with many climbers hanging out among the boulders, I made it to this hold for the first time. My forearms were melting and I hung for way longer than planned, panting and telling the encouraging onlookers that I'd been worried about how hard it would be to recover and get up Slap Happy from here. Glaswegian climbers are never afraid to frame their encouragement with a hard edge, and one of them shut down my excuses. 'Fuck's sake, Dave, how many times have you done Slap Happy?' He had a point. It was a party piece if you were strong, and I could climb it without slapping and then reverse back down. Following my pep talk, I dispatched Slap Happy and was amazed that a trick of timing the exact moment to turn my knee inwards as I executed the crux move, made in error because of my nerves and fatigue, worked so well that it felt easy. After trying these moves hundreds of times, the final secret revealed itself on the last move of the last attempt. Had I fallen and had to keep trying, perhaps I could have made further discoveries that would make it even easier. It was another lesson to be less afraid of new lines that seemed far too hard. It simply wasn't possible to judge the difficulty of the moves until you'd tried them for months, much less be confident whether they were beyond you or not.

With this in mind, I took a rope to the top of Requiem again. The crack in the lower half required fist, hand and finger jamming as its width fluctuated. But at half height, an eight-inch patina of rock had sheared off the whole upper part of the face, leaving a rail that ran horizontally on the left and vertically on the right. The second part of Requiem stepped right and laybacked up the vertical side. The crack continued just to the left but faded out to a thin, closed seam, no more than an impression without any obvious holds from a distance. A climb that followed the line of the crack all the way to the top was an obvious challenge, so I went for a look.

It was laughable. There were no completely smooth and blank sections, but aside from the odd tiny crimp, most of the features weren't really holds in any way I could envisage. I hung on the rope for ages, just staring at the headwall, trying to use my imagination and wondering if there was anything more I could really do apart from laugh and abseil back to the ground. Like meditatively returning to the breath from the distraction of the mind's chatter, I kept returning to the reason I'd come to this spot in the first place: searching for climbs that seemed impossible at first. On the boulders below, I'd already established that the question of whether something was possible

Opposite: Resting after the crux of Pressure (Font 8B) on the Dumbarton boulders.
© Hot Aches

could not be answered in a single session, or even ten. From my rope, I looked down and imagined the Requiem headwall as if it were a face on one of those boulders, at ground level. Then, I'd surely keep trying it. So what was the difference? If I were to take this climb, or my lessons from bouldering, seriously at all, I'd have to recognise that it would need to be worked out one move at a time. So why concern myself that its crux lay after 20 metres of hard climbing? That was a worry for another day. Just treat the last few metres as if it were a climb in itself.

All the same, I felt silly trying it, and preferred to go in the mornings when the place was quiet. I got around the dissonance by telling myself I'd just clean it and see if I could unlock a move or two, then leave it for someone in the future, as Andy Gallagher had done when bolting the walls either side of Requiem. Between sessions, as I drank tea on my windowsill seat, the headwall was always there, catching the sun and my attention. One day I started to daydream, the layout of its features in my head, when the phone rang. 'Hello?'

'Want a blowjob?' said the voice on the line. I looked up, confused. Then, a big fat man at the window in the tenement across the street caught my eye. He was standing naked, waving a phone (among other things) and laughing. *Fuck it. I'm going climbing.* Entertaining myself on that headwall might be silly, but it's all relative.

As was typical for Dumbarton, a sequence of one or two moves would run into a dead end. I could get my right hand to a three-finger edge, but it was only any use if I had my left hand on it. Snippets of sequences emerged like capillaries trying to connect to permit continuous flow. As sessions went by, I eliminated more and more dead ends, worked out more individual hand movements. At the same time, the 'stopper' moves emerged. There was no avoiding them, and once identified, I could dive deeper and deeper into the details, playing with the timing of weighting the holds and the amount of momentum to use. After ten sessions, spread over the spring of 2005, I managed every hand movement of the line in isolation. Somebody could climb this. I found this prospect exhilarating, even though there was no chance it would ever be me. In fact, with each nudge of progress, the whole exercise seemed less and less about me.

That thought snapped me back into a state of acute self-consciousness and then confusion about what I was doing, going down this rabbit hole on the headwall. The level of commitment, both in terms of the number of days and the physical and emotional effort expended, was greater than I'd ever given to a rock climb before, even though the idea of finishing it was too remote to be worth worrying about. So why make the investment? It slowly became clear that the curiosity itself was the reward. Without realising it, the difficulty of the moves and remoteness of the prospect of linking them

Top right: On the first ascent of You're Nicked (Font 7B) at Dumbarton, one of many unclimbed lines that remained on the Dumbarton boulders during the early 2000s.
© Claire MacLeod

Bottom right: Exploring new boulders around Arrochar with Niall McNair and Michael Tweedley in 2004. After finishing my master's degree, I hitched to Arrochar nearly every day over the summer, never waiting more than 15 minutes for a lift. I met many interesting people and climbed over 100 new boulders.
© Cubby Images

Right: Starting into the crux wall of Breathless (E10 7a) at Great Gable, Lake District during the third ascent. This route represented the cutting edge of trad climbing in 2005, with only a handful of E10 graded routes having been done. I repeated it in a couple of days, while still unable to do most of the moves on the Requiem headwall project.
© Steven Gordon

Right: Resting between attempts on the Requiem Headwall project.
© Cubby Images

Left: On the first ascent of Blackout (E6 6a) at Dumbarton. Grading new routes can be very difficult. This route seemed high in the grade, although amenable if you knew where the holds were. Still, I worried I should have given it E7 6b when Niall McNair opted to make an onsight attempt. He very impressively succeeded, although I found it a harrowing belay when Niall missed some important hidden holds. A fall from near the top would end on the ledges below the arête.
© Tim Morozzo

had shifted the question I was trying to answer from 'Is this move possible?' to 'How does this move work?' It was a subtle but critical difference. My own progress on the moves had become a side effect that had crept up on me all the faster for paying little attention to it. Even the physical effort seemed to wash over me. Unlike with training sessions on a board, it was not a cost I was counting.

When sore finger joints could take no more of the tiny crimps, I abseiled and wandered exhausted and hungry back across the wasteland, through the hole in the fence and round Morrisons, looking for the cheapest food I could find. I was woken from my daze by the shopper in front of me in the queue, inviting me to pass in front. 'Is that all yiv got, son? On ye go.' Dumbarton folk never let me wait with handheld shopping, even though I was a million miles away and in no hurry to be anywhere else as the crux sequence rolled continuously round my head.

It was already quite clear to me that a need to engage in challenging projects, climbing or otherwise, was part of my psyche. I had thought that this was all about seeing projects through to their completion by exercising discipline and resilience. That was rewarding, but this project seemed to reveal another layer. Unlike my route Achemine just to the left, it was not just about going forward, but going deeper. Claire and others in my family had often said that I had likely inherited or learned the tendency towards uncompromising curiosity from my mother. Like her, when faced with a complex issue, I became intensely focused on trying to understand it inside-out, sometimes losing sense of time and even perspective in the process. It was less a source of pleasure or entertainment, more a compulsion, and when this was harnessed by the practical skills of projecting a route, the combination was almost like magic. The moves on the Requiem headwall now done, I found myself starting to string them together.

I recalled that, with Achemine, the journey from completing the individual moves to linking the whole thing had been shorter than I anticipated. Initially perplexed by this, my hunch was that this was because the wall was only gently overhanging. On such small holds and with proportionally more weight on the feet compared to steeper terrain, even the most desperate individual moves linked together a little more readily. You could only use so much power on holds that small, so it took more of them to burn through your reserves. Could it be the same story on this project? This idea helped me to suspend disbelief a little longer and keep trying.

The Requiem crack, once my absolute limit, was feeling easier all the time and now only gave me a mild pump in my forearms. By the end of spring, to my amazement, I had climbed the line with one rest. It sounds good when I say it like that, but since

Top right: Getting shut down on the crux of Devastation Generation (8c) at Dumbuck in April 2005.

Bottom right: Climbing at Dumbuck in 2005, Dumbarton Rock behind.

the last ten moves were by far the hardest, I had really only climbed to the start of the difficulties. Nonetheless, I was delighted and eager to continue. The summer heat makes such hard moves impossible on a sea-level crag. So by the start of June, the pace of life slowed a bit and I had time to think.

During that spring I'd also been trying another new line at a nearby sport crag called Dumbuck. It had been my friend Dave Redpath's project, but he'd left it and said I could try it, so long as I used the name Devastation Generation for the climb, if I completed it. The name came from the early death of his father from a heart attack, another victim of the western diet and alcohol that defined 1980s city lifestyle in Scotland. The route was short and bouldery, and I felt it ought to suit me. But although I had milked my best asset to the limit and had the moves absolutely dialled in, I simply could not link through the crux to the top. The 'Groundhog Day' nature of sessions was getting to me. I knew precisely how each one would pan out: falling on the same move. A more fundamental change was needed and I began to interrogate every aspect of my training.

My favoured method, bouldering outdoors, will always be a core part of training for hard rock climbing. The combination of hard moves and attention to detail in movement technique is impossible to substitute. However, when trying 'real' climbs, the temptation is to hold off and attempt them in a rested state. Overdo this and you can miss out on the consistent intensity needed to gain strength, especially as you reach the upper end of the difficulty ladder. On many walks back from Dumbuck, I faced up to the fact that this had happened to me, and it was stalling my progress. But this was only half of the problem.

One morning in late June, I was sitting at home thinking and a more profound error in my approach finally became clear, and then, with hindsight, obvious. Complexity and intricacy of movement were what I had increasingly come to value and seek out in rock climbing. I could muster a big effort and pull hard on a crimp, for sure. But I was always focused on ways to minimise force and find every subtlety in a movement to get more weight on my feet. This is, of course, a good thing and something I'd cultivated into a powerful asset. Slowly but surely, my weakness had now become an overemphasis on movement technique, offering diminishing returns for further improvement. I needed to swing the pendulum the other way and get more basic. Summer was the perfect time to let go of short-term thinking and go back to building a foundation of raw strength.

Right: On the first ascent of Anubis (E8 6c) on Ben Nevis in 2005. I cleaned the route on abseil but decided to attempt it without practising the moves. I did climb it cleanly without falls, after multiple downclimbs. This was never my favoured style, but I greatly enjoyed the time exploring the north face of Ben Nevis in summer for the first time.
© Cubby Images

Right: Deadhang training at home on my £4 campus rung. Five sets crimp, five sets chisel grip, five sets openhand, six days a week. Consistency is the most important word in training.
© Claire MacLeod

13.

OFF THE RAILS

While living in Dumbarton, I'd followed up my undergraduate degree in Physiology with a master's in Medicine in Exercise Science, and I had been reflecting a lot on what I'd learned and could put back into my climbing. Before my studies, I had an image of a 'scientific' approach to training as embracing complexity and granular detail. As it turned out, the primary value of scientific training, at least for me, was in getting the basics right.

Like most climbers, I had initially gathered much of my training knowledge from fitness media, or from successful athletes. The fitness industry carries the constant burden of needing to sell either a fresh idea or a product, distorting its presentation of science. A single small research study overgeneralised to a whole population or an anecdote from a famous athlete often starts a major trend that lasts for years. The important is frequently eclipsed by the novel, unless it happens to be an old idea repackaged as a new one. Under the influence of fitness media, training culture can swing back and forth either side of an optimal path, constantly readjusting and overcompensating. Regularly bringing things back to first principles is a great way to cut through this and stay on the path. Overall, the core tenets of training are pretty basic, such as keeping it specific to the sport and increasing the load in a steady, consistent manner, giving the body time to adapt to the stress.

Simple as these principles are, human tendencies and aspects of modern society make them surprisingly hard to adhere to. In this regard, athletes can sometimes be their own worst enemies. They often show potential for success in sport precisely because they are difficult or even maladjusted people, prone to following their obsessions right off the rails. One common tendency is simply to overload volume and intensity. Indeed, training culture rewards the idea of athletes torturing themselves with a 'punishing' regime.

Much of the practice of designing training programmes and monitoring athletes is an exercise in holding them back from overdoing it. Of course, the goal is always to train as hard as possible, but the way to achieve this is not to pile on more and more load to breaking point, when it is too late to make adjustments before damage is done. Big errors like this are usually a hard enough lesson that they are made only once or twice, even by the diehard; it's the milder deviations that lead the athlete much further off track before the mistake becomes obvious. A training load that is slightly excessive allows you to just about keep up, with the comforting feeling that it will surely produce results. With no headroom to make the adaptation, over and above just absorbing the load, this actually results in less progress and a lot more frustration. At the time of my formal studies, the idea of 'readiness to train' hadn't properly crystallised in sports science discourse, and it remains an underappreciated concept.

The distance from my hard projects enforced by the summer heat made it easier to stand back and see more clearly how I should build this knowledge into my plan for the season. The way forward boiled down to a single strategy change and two tactical changes. The strategy was simply to abandon short-term thinking and approach training as if building a foundation, not a peak. I had to recognise that I would have to trust the process and neither expect nor test for results every week or even every month. Some of my earlier training had been aligned with this idea already, but in a messy and haphazard way, often because of injury. Instead, I would make it consistent. The tactics to achieve this were first to train in an environment disconnected from both actual climbing goals and competition with anyone else; and second, I would take greater care over the content of my rest time, so that I was ready for the next day. Putting as much effort into recovery would create more space to absorb the training and adapt. Taken together, these aspects ought to correct what I felt had switched from a strength to my key tactical weakness: constant focus on performing. I had become expert at making the very best of my ability, and now my ability was the bottleneck. Although the shift was subtle, it may be fair to say that the last week of June 2005, over a decade into my climbing life, was the first time I started training properly. I would begin with the most basic strength exercise of all.

Climbers had long practised 'deadhangs' on wooden fingerboards. But it was always seen, at least to me, as a simplistic or primitive training method, useful as an adjunct when 'real' climbing wasn't available. In fact, it wasn't in need of modernising; it should never have fallen out of fashion, and I followed the trend of renewed interest in it. Late as I was to experiment with fingerboarding, at least I made a good decision to commit to it for a whole season, long enough to give it a chance to work.

I had an old 20mm wooden rung from a campus board, and I screwed this into my door frame in the house and started to use it six days a week. In the past, I'd have spent my summers doing lots of trad climbing in the mountains, but this time I was in no rush to travel, and I only did a handful of routes, settling into a very simple routine and rarely leaving Dumbarton. I'd get up and do a little work and a lot of tea drinking, sitting on the window bay seat. Around noon, I'd do a few two-handed pull-ups on the fingerboard rung to warm up and then do five sets of hangs at my limit for each of the three main grip types: crimp, chisel and drag. The whole workout only took 30 minutes.

Straight after the fingerboard, I'd stroll round to the boulders to start my endurance work, which still needed to be done alongside the strength workouts. I'd base myself below a steep, hold-covered wall, away from the main crag and other climbers. I liked the quietude here, and I sat for long periods between sets, just looking down the Clyde and relaxing. I began to appreciate how this quiet, relaxing time, not thinking about anything in particular, seemed to help me recover from the strength work. I'd worked out two circuits on the steep wall. One was pretty easy. If it was really hot and the holds felt greasy, it was a little tiring, but I could still climb on it indefinitely. The other was a traverse I'd made up that was hard enough to fall off if I was really tired. I climbed in continuous loops for 40 minutes at a time, alternating between the easy and hard circuits, deciding which to follow while resting on a big jug at one end of the wall, above a big drop down to the shore below. The jug rest was essentially in a soloing position and, despite all the time I spent recovering there, my awareness of the drop below made it harder to relax my grip, which I liked.

One thing I'd seen when studying athletes, both in climbing and in my academic life, was that they existed on a continuum between those with explosive strength or a 'go all day' endurance phenotype. Often, choosing a sport that suited their phenotype at baseline would only potentiate the difference and settle athletes at the extremes of the range. Sports that demanded athletes sit right in the middle of these extremes, like 800-metre running, often had a reputation for being the hardest to train for. Climbing is unlike any other mainstream sport, requiring flitting back and forth between saving energy on easier terrain and explosive maximal effort on hard moves. Moreover, the switch often has to occur from second to second, multiple times—a difficult challenge to pull off. My default mode was to feel more at home on explosive bouldery moves. I lacked the ability to truly relax on easier terrain and thus save my power for the moves beyond. Studying muscle physiology helped me to see this as a whole motor program, almost a way of being that must be rewired over a long time if I wanted to shift away from my default tendency. As the summer drew on, I clocked

Left: First ascent of King Kong (Font 8A) at Dumbarton.
© Claire MacLeod

Above: On the first ascent of Drink Up for Tomorrow we Die (E7 6b) at Dunglas. I climbed the line while passing en route to visit my dad. He often used the expression, which neatly combined his romantic view of life and fatalistic grumbling about it over a pint. I tended towards reasoning over romanticism, but on routes as loose and dangerous as this, 'big talk' whispered quietly to myself was occasionally useful to keep me alive. The route has since fallen down.
© Barbara MacLeod

Top right: Another session working on the headwall.
© Steven Gordon

Opposite: Onsighting Kelpie (E6 6b) on Garbh Beinn, Ardgour.
© Cubby Images

Bottom right: At Dunglas with Claire.
© Barbara MacLeod

Opposite: Precious (Font 7C) in Glen Croe, Arrochar. © Tim Morozzo

up hour after hour hanging on the jug, holding more and more gently until it felt barely more effortful than standing on the ground.

Late in the evening, I'd go out for a relaxed run around the perimeter of Dumbarton, about nine miles. I stuck religiously to a very easy pace and primarily did it to wind down and empty my head of thoughts of climbing. It was too early to say if this way of training, new at least to me, might work. But it made me feel good in a way I had not experienced before. I'd never had so few rest days. Yet, as time passed, I still couldn't silence the niggling worry that I was undercooking the training since each week felt less taxing than the last. Between sitting at the window bay, long rests at the circuits wall and the shuffle round Dumbarton, every day contained three solid hours of apparently aimless daydreaming.

Each night around the same time I left for my gentle run, my neighbour let her dog out for its evening movement, unfortunately not the same kind of movement as me. At first, she accompanied it down the flights of stairs and out the back court to answer the call of nature. Soon, it was ejected out of the door on its own to shuffle down the stairs. Each week the dog was getting larger, slower and more miserable looking. It could no longer get back up the stairs, and I would find it still waiting to be carried up on my return from the run. Eventually, it could no longer even walk downstairs and simply opened its legs at the top of each flight and belly-surfed to the next landing. It was awful to see, in part because its decline was so rapid. Evidence of the damage done by a Western lifestyle, exacerbated by poverty, was ever-present in Dumbarton. I was away one evening when Claire heard screaming for help from the street below. A young woman living in the ground floor flat across the street had rung her sister to tell her she was about to take her own life. Her sister had rushed round and seen her hanging from the window frame. Claire ran to help and, after breaking the window, they cut her down. She was alive but did not have a good outlook.

Barely leaving the urban environment that summer, I appreciated more than ever that if things got in the way of spending time in the mountains or other natural spaces, it was even more important to move every day, as protection both from acute crises in life and the insidious effect of a Western lifestyle. Since moving to Dumbarton, I had seen a slightly different side to the cheery humility I admired in its locals - an occasional drift towards a fatalism that I did not share. I could get incredibly low, but this never led to apathy: I always wanted to take action of some sort, even the route of the woman in the ground floor flat. Climbing and the places it had taken me had gifted me a vision of a different life, which I was determined to pursue. Yet pointing myself at projects and putting a brick on the accelerator didn't work all that well either.

Like many keen athletes, I also found myself unable to maintain control and ending up somewhere I didn't want to be. Claire found my focus on climbing deeply frustrating at times, but losing myself completely in projects like the Requiem headwall felt so replenishing because when I did feel self-aware, I often hated myself. Like the empty cans of Special Brew I bagged and carried with me after my sessions, the dissociating effects of my escape into a world of sequences lingered long after I walked home from the rock. So did the side effects, almost destroying my relationship with Claire.

The summer had helped heal some of these problems. I had been heading in *almost* the right direction, a small nudge of the steering wheel preventing the crash that was imminent. I have been nudging it back and forth ever since. I ought to have realised sooner that sitting staring at the water flowing past between endurance laps was more than just waiting time; it facilitated the intensity I was looking for in life. Formal study of sports science didn't prevent me from making the same mistakes other athletes routinely make, but it did at least help me recognise when I had stumbled back onto the path I was trying to follow.

My only concern was that I noticed little difference in my physical prowess as autumn drew nearer. One of the few routes I climbed that summer was Wild Country on The Cobbler. Making the first onsight ascent of a route that had heavily influenced my initial desire for hard rock climbing turned out to be a big anti-climax. With no previous failures on the climb to compare to, I pulled over the top feeling that I had probably just got lucky and the crux holds I had slapped for were the right ones. On the fingerboard I had made unquestionable improvement, but only in the first six weeks; from the end of July until the start of October, the gains seemed barely measurable. I began to look forward to getting back to projects in the cold weather.

On the first cold day, I walked back over to Devastation Generation at Dumbuck to see if the crux felt any different. I pulled on at the second move, avoiding an awkward jump off a slab. On that first try, I climbed straight to the top and sat back on the rope at the anchor, wide-eyed in disbelief. In an instant, I knew my climbing had been transformed by the training—I just hadn't tested it in cold conditions when my hands weren't sweating. A few days later, I returned with a partner and completed the first 8c in Scotland.

Back in the spring, I'd been giving a talk about my climbing at Edinburgh University, and a friend, Paul Diffley, had chatted to me afterwards. He'd made an excellent film about himself and others climbing Scott's dry tooling route in Birnam Quarry, and he was keen to make more climbing films. He had asked if I had any projects, and I'd told him about the Requiem headwall, but that I wasn't expecting to be able to do it.

Right: First ascent of Devastation Generation (8c) at Dumbuck in 2005, after a summer of transformative training.
© Cubby Images

Left: Thinking about tactics while resting between attempts on the headwall project.
© Cubby Images

Nonetheless, Paul had said that he was keen to come and film me trying it, if only for the practice of shooting climbing. With the jump in my climbing standard, I got in touch with Paul and told him I would begin trying the line soon.

He joined me on the rope as I tried to link moves, asking questions to encourage me to explain what I was doing to camera. It was a quite different experience from the solitude I'd had previously on the headwall. When forced to vocalise the hurdles I was facing linking the moves, I analysed my sequence more quickly and decisively. Every time I chalked up my hands, I'd hear the beep of Paul's record button and feel a little pressure to actually make a few moves. Fortunately, I was ready to absorb and make use of this influence, and I found myself climbing from the ground, through the crux and then falling on the very last move, slapping with my left hand for the top of the crag. Slumping onto the rope, I looked incredulously at Paul through his lens. Everything had changed. This was no longer a project for the next generation. It was formally my project, and at that moment, I was locked in.

Keeping up the momentum with regular sessions on the route whenever it was dry, I soon reached the next big milestone of linking the whole route on the top rope. I knew this was coming, but when it happened, an intense sensation of dread washed over me, like I had fallen into a fast-flowing river and my direction was now entirely captured by the project. Stepping away from it now was unthinkable. It had fascinated me as a curiosity, but it had gone past that. I was now part of it, whether I wanted to be or not. I could see other things happening in my life, but they would slip past at a distance as I fought to stay afloat and get to wherever this might lead. The routine I had set for myself over the summer was long gone, and I anxiously watched the forecast that would now dictate my schedule.

Like Achemine, the headwall project had good protection but a long, scary runout all the way to the top. A fall was most likely from a small undercut block about eight moves from the top. If I got through that move, I'd usually make it to the last move and fall there instead. The line was slightly contrived in that, three moves from the top, it was possible to take an edge with the other hand and slap out left to a different finish. The final three moves avoided by this variation were not that difficult in isolation, but with power fading at the end of the route, I consistently fell there on most attempts. Tempting as it was to take this line, which might still be Scotland's first E10, it would just be avoiding the challenge. The entire concept of the climb was to take an uncompromising approach. In practice, that meant a very high probability I'd be taking 60- or 70-foot falls from that last move.

I didn't have any excuses not to attempt a lead. No trad route existed with this level of

Opposite: First ascent of Apollo (8a+) at The Kracken near Tighnabruaich. © Michael Tweedley

difficulty, so further practice or training to try to add a margin of fitness was unrealistic. I was already at a peak of fitness higher than I might have dreamed of. The falls would be intimidating, but ultimately, I wasn't going to hit the ground. I worried about the undercut move lower down, though. It involved stepping my foot awkwardly across my body, and as I did so, a trailing lead rope would end up behind my leg. My foot often slipped at this moment. If that happened on lead, I'd flip upside down and potentially get tangled in the rope. I couldn't think of a way to avoid this scenario other than not to fall off this move. There was nothing more I could do to mitigate this possibility. I'd just have to go with it. As with all climbs, when the details were all worked out and I had done everything I could, it came down to a simple decision about how much I wanted to lead it. Was there enough fear to walk away? Or would it be overcome by the curiosity? This decision was just creating an illusion that I could still choose. With so much work done on the curiosity side, fear would now have an impossible job of catching up. Even the timing wasn't really under my control any more. It would be the first day cooler than 14 degrees with a strong westerly.

It soon arrived, and I tied into my lead rope and started up the Requiem crack.

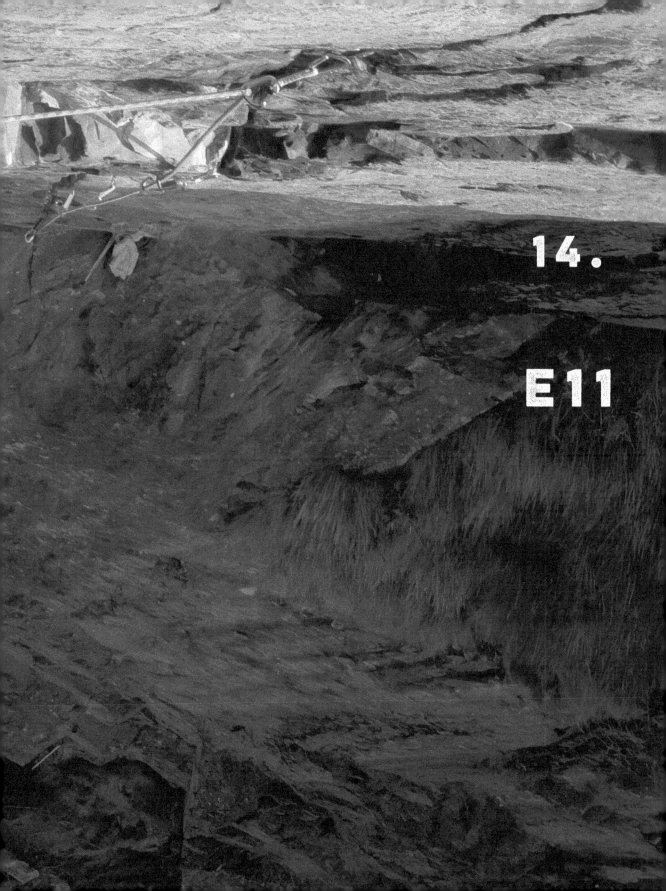

Left: First lead attempt on Rhapsody (E11 7a), moving right to enter the crux section. It is crucial to get the left ring finger to bite correctly when matching the tiny edges.
© Cubby Images

Previous: The initial moves away from the jug rest at the foot of the headwall.
© Steven Gordon

After such a long time rehearsing moves with the safety of the top rope, the prospect of leading the Requiem headwall brought a polarised mix of emotions which weighed on me day and night before my first attempt. I was massively intimidated, knowing that the probability of success was very low and so a huge fall was virtually guaranteed. But it was exhilarating that all the practice, anticipation and agonising over details were now coming together on the sharp end. This was real.

As I worked my way up the crack, filling its deep slots with cams every few metres, tackling the attempt section by section helped keep a lid on the dread of launching up the headwall above. Instead of the looming crux, I focused on the sound of my feet scuffing in familiar jams and my sharp breaths between movements, still controlled but fringed with jitters. This worked well and the crack flew past, without errors or delays. After placing the top runner where the Requiem crack faded, I stepped left to a big flat jug and hung from it, left heel by my hand. The imminence of the outcome was buzzing in my brain like a siren. I knew I had no more than two minutes to try to recover both strength and composure before tiredness would slowly creep in. This should be a simple task, just waiting, swapping hands, letting the concerns of the previous section drift into the past. Yet I couldn't get comfortable, and it was very hard to relax on the jug, to hang with just enough force not to let go. The headwall was the hardest sequence of moves I'd ever linked, and anything less than full recovery would reduce my chances of getting to the top close to zero. To truly relax, I'd need to uncouple the tightness of my grip from the anticipation of what I was about to do. But I could not.

Every instinct - fear, anxiety, pressure - demanded that I tighten up, even though it was essential to hold back that energy until the traverse rightwards into the crux, ten moves beyond. I fought to block it for another few minutes yet, stemming its leak into my body by focusing on the friction of the storm-washed basalt rail under my fingers, the cool wind carrying away my body heat, and my breathing, now finally losing the jitter on its edge. At some point in the second minute, there would be a nadir in this forced relaxation, and the discomfort of my strained position with one foot above my head would steal the last per cent of energy I'd need for the final moves. I listened intently for this low point. It would dictate the exact moment to launch. Tension waxed and waned on each breath as I swapped hands back and forth. Then, on two or three breaths in a row, it fell no further. It was time to move and release the pressure.

As I had hoped, the focus of executing the first moves up the headwall snapped my attention back to a familiar programme of big spans between flat sidepulls. This section, only 8b, had to fly past without any strain, otherwise the attempt would be over before it started. It was best to let muscle memory do the work while my mind

This page: First lead attempt on Rhapsody, fingers about to uncurl from the undercut block.
© Cubby Images

got ready to pull the pin on the tension beneath at the correct moment. I landed on the two positive edges before the short traverse right that marked the entry to the crux. I lingered here for ten seconds, feeling the first signs of fatigue in my forearms and the imperative to move again. There was no choice but to reach the tiny edge out right and attack the crux. Bringing the left hand in to match, it was critical to get my ring finger to bite on a little bump in the otherwise sloping hold. It took a second or two to find the spot, and I had a moment of thinking space to marvel at the fact that I was really here, on the lead on a next-level project. That thought was overridden by another, more urgent concern: my arms were feeling it. I knew what was coming, and as my fingertips flicked off the undercut, I was flying.

The rope whipped and spiralled, suspended in the air beside me as I dropped, recalling the memories of the 11 falls I'd taken on Achemine. That snapshot disappeared in a blur of acceleration, the yank of the rope and finally the wall coming back into focus, straight towards me. With a loud smack, my feet slapped hard off the rock and it was over, everything quiet as I spun around in space. I lowered down to process the attempt. It had been a shaky one, with a few inefficiencies and tiny errors that always accompany anything unfamiliar. I had probably only added a second or two to the climbing time with little foot adjustments to take account of the lead rope. For a future attempt, I'd have to iron them out, but that didn't trouble me. What was troubling was that the easiest part of the entire pitch - the rest at the foot of the headwall - had felt so different on lead, and I didn't know why.

As soon as I'd linked the climb on the top rope and said to Paul that I was going to fully commit to trying to lead it, he said he wanted to film every attempt. Given the frequently rainy October weather in Scotland, I had no option but to give him at most a day's notice. Showery days were often the best, with a fresh wind blowing up the River Clyde. I'd ring Paul to say that tomorrow was a maybe, and then he'd have to take the day off work, even if it ended up rainier than forecast and the wall stayed wet. He had to invent an unpredictable health problem to justify taking so much time off, and opted for irritable bowel syndrome - a great choice. Whenever I'd ring Paul to tell him to come to Dumbarton the next day, I'd have an immediate bout of irritable bowel syndrome myself. Mine was more predictable, though: I still worried about my leg behind that rope.

On the next lead day, Paul came round, and while a westerly gale battered Dumbarton with black rainstorms in regular salvos, we drank tea and I nervously paced and peered at the Requiem wall from my window. The front soon cleared as forecast, leaving just the raging gale, which turned the wall from black to pale orange in minutes. Time for action.

We were soon going through our preparations at the crag, Paul making his way to his perch at the top of the face. I felt sorry for him, getting flung around in the gusts and struggling to operate his camera with an ungloved hand. As I warmed up, I could tell the conditions were amazing: the rock seemed like Velcro. I'd never felt it this sticky, as if every pore and crystal had been pressure-washed by the morning storms. Climbing the crack, I thought of past days when I'd turned up at a crag and found the rock friction exceptionally good. In these circumstances, moves feel noticeably easier, sometimes even completely different. Yet the advantage can be offset by a sense that you ought to succeed, and you become nervous and jittery; errors creep in and sabotage the opportunity. I'd gradually learned not to fight against, and thus intensify this pressure, but instead to just let it wash over me. By accepting the inevitability that your brain will produce these thoughts, their potency is diminished, letting you refocus on what you can control: your moves.

I was still bouncing this idea around in my mind when I registered that I'd passed my previous high point and my arm was wrenching into a deep lock on the next sidepull, six moves from the top. As I pulled up, the contrast with my detached state from a moment before could hardly have been more profound. In an instant, reality crashed

Opposite: Reaching left for the weathered credit card sidepull four moves from the top. Dumbarton basalt is often polished and slippery, but the combination of recent rain, sun and cold wind can create friction that makes even tiny holds feel secure. © Steven Gordon

back into my consciousness like a punch in the face. I could feel the weight of the rope and the long belly of slack as it swung in the wind between me and the micro-wire far below. The tiny holds still felt grippy in that wind. But with fatigue, their bite was fading into numbness. Nonetheless, I made it through the hardest move and to a stretch left for a lovely, weathered sidepull, minuscule but incut, sticking out of the rock like the edge of a credit card. It's amazing that it didn't just snap off. Although only a few millimetres deep, the sharp edge digging into my fingertips gave a brief glimmer of security as the power in my arms faded.

The cold wind, the bite of the holds, Paul's muffled shouts of encouragement from behind his camera and my heart thumping in my throat all felt hyper-real. The most real thing of all was my committed position. *I can't be here. I am here. I'm about to fall. The strength isn't there for the last three moves.* How to respond to thoughts like this while climbing depends entirely on the exact scenario. On routes with the potential for getting badly hurt or killed in a fall, you really shouldn't be in a situation either where things get out of control or fear takes over - or both. It's imperative to prepare a strategy for maintaining control over the climbing and your feelings before you even leave the ground, or before you reach a point of commitment from which you cannot retreat.

On climbs like this project at Dumbarton, where the fear is merely of a long fall and not of dying, things are rather different. Although the consequence of a fall is less serious, the probability of it is extremely high. This proximity to the prospect of falling is likely why many climbers often become excessively fearful on relatively safe climbs. Fear of falling is real, neither easy to ignore or overcome, and it can be crippling for even the most confident person without the right training. Again, it is better not to wait until the actual lead before dealing with it. I had spent over 20 days trying the project on the safety of a top rope, and after each session, I spent the walk home and many more hours lying in bed visualising, adding a layer of fear on top of my memory of the moves to test how this felt. I let my mind drift to the extremes of how crippling the fear would feel. My heart would pound, my fingers would sweat, and the nervous energy would make me squirm. Despite the temptation to think about something else, I'd let the rehearsal run on, over and over again, exposing my mind fully to the discomfort, almost revelling in it. Still, this pale imitation of the real thing could only take the edge off the fear; it couldn't come close to eliminating it. Fear of falling can be chipped away at with tactical knowledge and visualisation to gain familiarity, but the bulk of it may remain, still in your way. The heavy lifting of overcoming this is achieved with a far more primitive clash of arguments, one of which must ultimately come to dominate over the other.

The only thing I have ever found more powerful than the fear of falling is the desire to complete the climb. Not necessarily *this* one in particular, but hard climbing in general. Would I even want to eliminate fear completely? It serves the rather important purpose of lowering the probability of getting killed during a lifetime of trying these types of routes. If I wanted to fill my life with increasingly hard climbs, I was going to experience this feeling a lot. Even if I could silence fear on today's project, tomorrow's project would see its return. I would need to live with its presence somehow, perhaps even find a way to turn it into an attribute.

Repeatedly failing on hard, bold climbs had shown me how to manage this. If I backed off, grabbed a runner or simply fell, the acute feeling of fear died as soon as the attempt was over. It was immediately replaced by the discomfort of the failure itself. Since my solo on The Cobbler as a schoolboy, I knew I could not rid myself of that discomfort. Unlike fear, it would not fade with time but instead deepen. There was one exception: if I had done everything possible to change the outcome, turned over every stone, explored every possibility and it still didn't work, that would be fine. I'd never yet come across this situation. All my experience told me that the rock yielded to improvements in movement sequences even after years spent trying routes. Moreover, the basis of the enjoyment I got from climbing was the search for those improvements and the bitter taste of failure no longer registered as unpleasant. That acquired taste for discomfort could also overpower the fear of falling, and sometimes cause it to flip from a siren for retreat to a driver for ratcheting up intensity of effort.

As I pushed on towards the final few moves on the Requiem headwall, I just couldn't bear the thought of not finding out how much further I could get if I gave it total commitment. What a sickening waste it would be to reach this point, then half-ass the opportunity because the fear felt a little intense. No. Let's put all the cards on the table and see how much higher I can get.

Not much higher, it turned out. As I reached again with my left hand to a tiny sloping edge, I had the dreaded 'that's not even a hold' feeling you get when you don't have anything left in your forearms. The edge has a ripple, and normally I'd feel my fingertips bite on it. I bounced my fingers repeatedly, searching for the bite. Nothing. I still had enough left to hang on and bring my right hand up to a sloping pinch, but as I took the hold, I could feel my strength vanish. The next move was the final one. All that was needed was to lift my right foot, pull up and make a straightforward lunge to a big flat hold on the top of the crag. I could see the grass peeking over the top, right above my head. With the last of my power, I pulled and slapped, but it wasn't enough. By the time I got my left hand to the jug, my body was already parting company with the wall. I was off.

Below: Mid-fall after coming off the final move. As the rope came tight, the top wire snapped, and I eventually stopped level with the slab below the halfway ledge.
© Steven Gordon

It was going to be a big one. I was resigned to it, and so I just took in the awesome sight of the arcs of rope floating in the air as I sailed past in a blur. I expected to stop around level with the belay ledge at the start of the Requiem crack, 60 feet below. But the rope whipped tight, then released me again, only to catch me a moment later in a more gentle swing. I was shocked to recognise the holds on the easy initial slab in front of me. The top micro-wire had snapped, dropping me a further ten feet but softening the catch. Mark Garthwaite, who had been walking round the path to the boulders just in time to witness my flight, shouted up, 'They don't get much bigger than that, boy!' I whooped in agreement. The softness of the fall and Garth's dark laughter helped to evaporate the struggle and fear of the previous moment. The aftertaste was an intense, burning sensation that I had done something really important. I'd used fear to ratchet up commitment and effort enough to squeeze out six more of the seven

moves above the previous high point, literally touching the possibility of success, with the top hold in my hand. That felt tremendous.

Naturally, conditions weren't quite so exceptional on the next dry day. The novelty of the first couple of tries had also worn off, leaving only the sense that all the work was still to be done and it wasn't feeling any easier. I took two more falls from lower down, trying to move my feet off that undercut. On the third and final attempt of the day, my worst fear was realised. As I crossed my foot behind the rope and stood up, it slipped suddenly, and I plummeted. I only had time to register the spiralling rope wrapping twice around my extended leg, crushing my calf and burning my skin as it came tight. I flipped violently upside down and battered off the wall with a scream of fright and pain. Wincing, I unwrapped the rope from my lower leg, which was trussed like a roasting joint, and asked to be lowered slowly down.

Afterwards, I was okay, with just a very bruised and sore calf. But the buoyant feeling that I was making progress had gone - the climb had bitten back with a vengeance. I returned to the top rope to try and improve my sequence on the step-through move only to run into a much bigger problem - the little undercut block needed for the crux move broke off in my hand. In an instant, it was a completely different route and perhaps not possible at all. After a couple of hours, I worked out another method, but it was harder and added more moves. I'd need to go back several steps and re-link the moves on the top rope again. My 'last move' high point meant nothing now.

With the late October rain intensifying, the window of opportunity was gone, although I was in denial about this. I still sat in the house, searching the forecast for dry days. All the while, the winter seepage on the wall was setting in. Late one evening in mid-November, I was walking through Edinburgh with Paul after speaking at a climbing festival. I raved to him about a couple of drier days forecast the following week and how there could be more chances to try the route before Christmas. Paul nodded dismissively and kept walking. Perhaps it was just that it was late on a cold, wet walk across town, but it struck me that my optimism seemed ridiculous to him, barely even worth a nod of acknowledgement. I stopped talking, but continued to churn it over inside my head. Sure, I must have sounded like I was clutching at straws, but how else would I actually make progress? I felt I had to grab every straw. That discomfort I was looking for tasted as sharp as ever. I was so mentally locked in to the project that I was completely blind to the facts in front of me: that it was over for the year. With hindsight, perhaps Paul didn't want to hear any more about this project because he was in danger of losing his job. Too many days of irritable bowel syndrome, not enough getting to the top.

Right: Crossing my right foot across the rope on the crux. With both hands on very poor undercut holds, there is a force greater than body weight going through the foothold which is about 2mm deep. The foot often slipped, and I was scared I wouldn't be able to get my leg clear of the rope during the fall.
© Steven Gordon

Through the winter, I couldn't detach and felt continually keyed up and preoccupied. I didn't make home life for Claire easy. Justifiably, she felt that my focus was lost to the project and she stopped trying to encourage me to think about anything else. I totally failed to grasp this and spun my wheels on boulder projects at Dumbarton, finally climbing the line out of the black cave, which I called Pressure. With my foot still on the accelerator, the physical pressure was undoubtedly making my body stronger, which was necessary to convert the 'nearly' attempts into success, especially now the route was harder. But that steady balance of preparation and discipline to rest as well as train that I'd had the previous summer, when progress had come easily, was gone - I'd veered well off the path. Only the frequent rain saved me from overtraining. It seemed like the moment the headwall project had become hard was the day I linked it on the top rope and it switched from being a thing of pure curiosity to a real goal.

Right: A sign of improved fitness, able to chalk up before taking the sidepull.
© Cubby Images

Still, I couldn't see the difference: I was training just as hard as I had that summer, and now that I'd come so close to completing the project, I had a much clearer idea of how I wanted the future to unfold.

March couldn't come quick enough. As soon as the afternoon sun lingered in the sky long enough to hit the Requiem face, I was back on the wall, alone. With the weaving nature of the climbing on the headwall, it was difficult to do long links with a self-belay setup, and early in the season, my usual partners weren't around. But after Paul's reaction to my over-optimism, I wanted my next call to him to carry news of real progress. Claire offered to belay me on the top rope, and in her company, I linked the moves again, first try. She had belayed me on so many hard and serious trad routes, but never this one, and we joked again that all that was needed was Claire on the other end of the rope. The first sun of the season on my face seemed to recreate the vibe of the previous summer.

It was time to ring Paul. Next day, in the faint warmth of March sunshine, I led the Requiem crack, placing the runners, before downclimbing all the way to the ground again. This now felt so easy that I had to continue the warm-up afterwards on the boulders below. Paul resumed his hanging vigil two metres right of the top of the project and I asked Michael Tweedley to belay from the ledge at the foot of the crack rather than the ground, so I'd have a little less rope to trail to that last move. Feeling so much stronger during the session with Claire, it seemed like the right moment to pile on every marginal advantage I could think of. After climbing the lower slab, I shouted up to Paul that conditions were excellent and went for it. Through every move on the crux, I could feel the winter of training in my arms. Reaching left to the sharp sidepull, I could even take a moment to flick my wrist and get some recovery in my forearm before curling my fingernails into the back of the hold. It felt like it was on. Yet once again, as I pulled up to make the final slap to the top hold, I had nothing left to give. My left hand touched and then released from the top of the crag, and I was looking down through my feet at the space I was falling into.

Lunging hard for the top hold had pushed me outwards from the wall, so the swing back in as the rope came tight ended with a painful smash. Tension that had been simmering inside me since October broke free and I screamed out in frustration, 'WHAT THE FUCK DO I HAVE TO DO?!' The soles of my feet had taken a good whack, and I limped off home with the question still repeating in my head. I couldn't take this beating anymore, I thought, mistakenly.

After a night's sleep, the curiosity came right back, and with it, my answer: I had to go back to the sequence. On the route's hardest sections, I had invested huge

Previous: I was fitter and stronger, but it still wasn't enough. © Cubby Images

amounts of time working out the fine details of the handholds and footholds. I had not done the same on the very last move. In isolation, it wasn't hard. On endurance routes like this, however, it's not always possible to know which of the easier moves will end up becoming stopper moves when you are fatigued from linking the climbing all the way from the ground. I needed to return to that last move and see if I could unlock anything further.

Still tired and sore, I limped round to the crag with a spring in my step. Experience told me there would be something there if I looked for it. I lowered down two feet from the top and stared at the rock. After some experimentation, I found a tiny nick for my left foot. When slapping for the top of the crag in isolation, it made a barely perceptible difference. Had it been on a boulder, I'd never have bothered with such a tiny detail on an easy move. But it did seem to be just a little more stable, preventing the left side of my body from sagging so much and turning a lunge into a snatch. I walked back home feeling the same glow of satisfaction I always did after being at Dumbarton Rock, though I acknowledged that it did feel a little obsessive to go to the crag with the intention of doing only one move. I'd have no idea if it would make any difference to the overall outcome until I tried the whole link again, and as I rang Paul to arrange the next day at the Rock, I decided this would be another lead attempt.

Paul had been bringing more and more camera kit to the crag each time, and we lugged large protective camera cases round, creating a big mess of kit below the route. No sooner had we arrived than a heavy snowstorm turned the place white in seconds. We scrambled to pile cameras and bags underneath boulders, the kit, the crag and ourselves soaking wet. It was a farce, one of those days where the climbing looks over before it's begun. The mood lightened, or at least mine did, and I started to play on a roof project on the boulder we were sheltering under. It was a shame the day was a write-off, as I felt strong and the holds felt sticky.

Once the storm passed, the sky cleared and gave way to rich evening sunlight and a dry, freezing wind - an air that is difficult to describe but only seems to occur in early April and often coincided with days I had completed hard new routes. With the sun now on the Requiem wall, I started to wonder if perhaps it might be dry enough for an attempt before sunset. I quickly rang Claire and asked if she might come round to belay me just in case, and she appeared soon afterwards.

Claire looked on silently as I scurried back and forth among the mess of kit below the face, arranging cams and laying out ropes as the light turned orange. This quietness was a familiar and welcome space within which to go through the process of readying for a huge effort. There was no need to offer encouragement. Both of us knew that I would give everything to the attempt. We also knew, at this point, that I could do the route, but importantly, that any single attempt was still highly likely to end in another massive fall that Claire would shortly be fielding. The quietness was a signal of understanding. Claire knew every weakness I had. If I was at the point of tying into the rope and there was nothing else to say, then I was ready.

I knew of many great partnerships in climbing, some personally and some by reputation. Climbers often spoke of how their partner was able to extend their own abilities beyond sharing the workload or gathering the knowledge required to complete a hard first ascent. It actually seemed to make both of them more capable in ways that were difficult to pin down. Finding these partnerships requires good fortune, and although I had many fantastic climbing partners, few overlapped closely enough with the same set of disciplines I wanted to excel in. I'd spent my youth chasing Dave Cuthbertson's routes, and while our motivation overlapped almost completely, he was forced out of climbing by injury just as I started to gain the skills to look to the same projects he had. Although she did not climb, Claire had been the person I'd come closest to having this experience with. Where I might otherwise spin in circles of thought for months in an attempt to figure out how to approach a climb, Claire could perfectly reflect my challenges back to me with nothing more than a look, and in so doing, remove all the noise from my head.

I reached the halfway ledge and sat, still enjoying the quiet, looking out to the Clyde, down at Claire and up at Paul in his usual position hunched on the rope on the skyline. Despite everyone waiting for me, a feeling of comfort washed through me. The polarised mixture of dread and anticipation that I'd felt on this ledge on each previous attempt had gone. I just felt at home. Everything felt straightforward and normal. I was weak. I wasn't half as good or strong as Cubby, who had been here on Requiem in 1983. I was probably going to fall, again. It was now April and I'd soon lose another window of conditions. I'd probably be worrying about this tomorrow, even though it wasn't within my control. All the usual thoughts that preceded my attempts were running their course. But somehow they seemed quieter and inconsequential. Like the sounds of Dumbarton life, they carried on in the background while I waited for my moment to do what I was here to do. Discussions of moves in Glaswegian accents floated up from the boulders below, and in the distance, the six o'clock train clacked across town, taking folk home from work in Glasgow to make the tea and watch telly.

The sun edged lower, just above the Cowal hills, dictating that I must stand up now and begin. I called gently down to Claire that I would climb and then I started up the

Right: Calling to Claire from the halfway ledge, ready to start my attempt.
© Hot Aches

crack. At the base of the headwall, the rest position was its usual uncomfortable self. But the friction on the flat jug made it feel good to be here, craning my neck up at Paul. The pads of my calloused fingers dragging off the sharp edge took all the load, forearm muscles stretched but soft. It reminded me of summer on my circuit below and my breathing dropped almost to nothing.

Some attempts on hard climbs feel effortless and flowing, sometimes like an out-of-body experience. Others are very much in the moment - the bite of sharp, painful holds, the grunt of muscling through moves, the taste of fear on a dry mouth. Moving through the crux, I noted that this attempt felt like none of these. Although I was on the edge of falling on each move, the constant uncertainty did not trigger the spiral of thoughts it usually did. The quietness I had felt on the ground still filled my consciousness. One less 'bounce' adjustment was needed on the left-hand crimp, two moves from the top. That was the first clue that something was different. Paul saw it. The tone of his mantra 'Go on' kicked up a note: 'Go ON!' Then I reached the last move and pulled up, with enough strength to aim my left toe and wrap it carefully onto the little nick I'd found the day before. That was the next clue. 'GO ON!' Another note higher from Paul as he saw what was happening. With a quiet 'whack', I heard my left hand land on the top of the crag. And it didn't let go. The hold felt huge, ice cold and grippy. 'GO ON, YOU

Right: Claire relieved to see the project complete.
© Hot Aches

FUCKING BEAUTY!' yelled Paul from behind his lens as I flowed over onto the ledge.

Rocking up onto my feet, I couldn't stop moving forward and I raced across the ledge, slapping my hands on the rocks at the back with a scream. I didn't know where to put myself and I paced around, hands on my head in disbelief. On the wind, I could hear clapping from the climbers who had stopped bouldering to watch the attempt, no doubt expecting another 60-footer as much as I was. Their whoops of delight seemed more than just polite congratulation.

Paul and I scrambled to the top of Dumbarton Rock and stood on its summit, looking across to the Highland hills. As the sky turned purple over the snow-covered mountains, the sting of the wind on my face added to the sense of hyper-alertness as I buzzed with exhilaration. Paul pointed his camera and asked me to make sense of the climb. I had barely even registered that the route was really done, never mind fully grasped that it was the hardest trad route in the world and would be the first to earn the grade of E11. I had thought of nothing else beyond the moves for months. It was hard for me to even remember when I entered the tunnel of focus. Right now, I felt blinded by the light at its end.

As the last rays of sun faded on the summit of Dumbarton Rock, Paul and I began to shiver, and it was time to abseil down. Paul asked one more question of me: What

Right: Packing up my harness and a chapter in my life, excited to start the next one.
© Hot Aches

would I do next? It was strange that I had got as far as having done the hardest climb of its type in the world, at the very same cliff I started on, just along the road from my house. My climbing apprenticeship, although deep, was not yet very broad. I told Paul that I didn't know exactly, but I was looking forward to wherever climbing would take me. The uncertainty was an appealing thought, and already the celebration of climbing Rhapsody was starting to fade.

EPILOGUE

Left: First ascent of Metalcore (8c+) at The Anvil, Loch Goil, May 2007. © Claire MacLeod

Later, I would feel awkward about the fact that I had been the first to climb a route of this ridiculous grade, and acutely conscious that others would wonder how it could happen given my lack of natural strength and talent for movement compared to other climbers of my own and, in fact, previous generations. My strengths may not be as obvious as doing a one-arm pull-up on a tiny crimp, and this is why I have laid them out in this book. I brought to my very first day of climbing at Dumbarton Rock a willingness to fail and an absence of any fear of failure. Some may see this as a lack of self-confidence, and I certainly did at the time. I now see it as the reverse. As I progressed through my climbing apprenticeship, trying to find my limits and observing my peers do the same, I could see that constant self-reinforcement and belief in the capacity for or inevitability of success could be a weakness that eventually thwarted progress. It can get in the way of solving the problems that must be dealt with as you progress, often by preventing you from noticing those problems in the first place. Acceptance of the possibility that any project, climbing or otherwise, might not work, tends to draw your focus back to what's important: figuring out *why* it might not work. I now think accepting the possibility of failure *requires* self-confidence.

In a sort of Hawthorne effect, where the act of observing behaviour causes it to change, seeing and accepting your limits can be the first real step to pushing them back. In a sense, my early approach to sport took the potential for worry about success or failure and simply moved it out of the way, allowing me to focus on learning to climb. My education in sports science helped me see that tweaks in training methodology tend to produce lacklustre or even null results. The big gains come from correcting more basic mistakes. The more I learned about training, the less I worried about the details and the more I looked after the foundations. Despite my formal education in the field, I was rarely able to anticipate adjustments I needed to make but did at least

Right: Starting up Sanction (Font 8B) at Dumbarton in 2007. I was very close to climbing an extended version of this, starting in the cave, but decided to leave it behind when I moved to the Highlands in June 2007. It was climbed the following year by Malcolm Smith to give Gutbuster (Font 8B+). It remained unrepeated until I returned in 2019 and made the second ascent. © Claire MacLeod

recognise that deep self criticism could cut through heavily engrained habits of both thought and behaviour.

Of the mistakes I can probably take credit for noticing and correcting, simplifying my life is likely the most important. In practical terms, this meant removing hurdles to accessing training and making space in my life to undertake it by lowering the cost of living. The tactics to achieve these aims were to base myself ten minutes from the hardest crag in Scotland and to hang a 20mm rung in my doorway. Those things had upfront costs, which I am grateful I could meet: £80 per month for the house and £4 for the wooden rung. But after that, Dumbarton Rock and my rung were free to use daily. To this day, I believe that the single most important training decision any climber can make is to live very close to the right training environment.

Other corrections I cannot take credit for, stumbling into them by accident and sometimes having to relearn the same lessons more than once. Although I'd learned to pay attention to my state of readiness to train and take proper rest days, as my resilience to training load rose over time, it was easy to forget this, especially when

Right: Ready to move on to a wider horizon of crags in May 2007.
© Claire MacLeod

deep in the process of projecting hard routes. It was only with hindsight that I could see the immense value of my 'steady' routine of training in the summer of 2005, detaching from short-term climbing goals for months on end and taking time between endurance circuits at Dumbarton to just sit in the grass and daydream, finally giving my parasympathetic nervous system a chance to get to work and my sympathetic system a moment of respite.

Nor can I take credit for the single biggest advantage I have as a climber: curiosity. I first noticed its burn after backing off North Wall Groove on The Cobbler. At the time, I mistakenly interpreted it as frustration with failure. Later, I came to appreciate it in the same way I do the physical pain of burning muscles during exercise and training. The sensation fluctuates between an ache that is pleasurable, even addictive, but occasionally ratchets up to become unbearable, demanding all my attention. At this level, curiosity becomes immensely powerful, even dangerous. It can drive someone like me, with unremarkable levels of discipline, to sustain much higher levels of effort or tolerance of fear, pain and uncertainty than I ever could without it. I think that I have

Right: With my mum, Barbara MacLeod, in Glen Nevis, June 2007. © Claire MacLeod

inherited, or learned, most of my curiosity from my mother. Despite being handed this huge advantage before I even made a single move on a rock face, I do, however, believe that it is a trainable skill. Anyone ought to be able to transform a playful inquisitiveness to answer a question like 'How would this move work?' into a burning need.

So why does this not happen more often? For me, the process of being drawn in to answering curiosity feels like a gravitational pull. As we drift into new experiences in life, we pass across the outer orbit of many different questions. The strength of pull will determine whether we continue onwards, begin a loose orbit, or are drawn into a spiral of ever stronger engagement. My strong sense of curiosity has defined my life as a series of violent spirals towards climbing projects; bright stars or black holes, depending on your perspective. Claire has described my lifestyle as lurching from one project to another. I find being caught in the spiral exhilarating, and after completing projects throws me off in a new direction, the quiet drift is a welcome contrast. However, if I drift for too long, I find that the space becomes lonely and dark.

I do believe that others can, if they wish, place themselves deeper into the orbit of questions by actively adopting habits that alter their direction. Choosing to closely observe the details of missing information, sharply defining the edges, and tolerating the discomfort that arises from not understanding something, tends to produce motivation. This, in turn, has the power to skew your direction much further into the orbit of the question, and escape becomes impossible. At root, I consider my own learned attributes in climbing to be keen observation of details and tolerance of the discomfort of the gaps it exposes. Once these habits have done their work and I'm caught in the spiral, the pull of the question does the rest for me. The only thing that requires real effort on my part is to stay in one piece while I'm thrown around.

Having decided that climbing was my desired activity, I had the good fortune of stumbling into an early goal to climb the hardest route in the country. It was an accident that the question I became curious about was, 'How hard can I climb?' I'm not sure I'll ever have the answer. There is always more to learn about how to move on rock.

Shortly after climbing Rhapsody, I began a period which is more synonymous with being a professional athlete than the routine I have described in this book. My life became more complicated with more opportunities to share what I had learned through writing, filmmaking and speaking about climbing. Much as I miss the simplicity of the learning period during my apprenticeship, I have continued this less focused lifestyle to this day, since it brings its own rewards. The first step in trying to give back to the climbing community to which I owe so much was starting to write a blog later in 2006. It became popular, and readers told me they appreciated that I was always trying to show my working and deconstruct how I was applying the principles of training, or learning the hard way, in real time.

Later, I crystallised these fundamentals of training and tactics for hard climbing into my first book, *9 out of 10 climbers make the same mistakes*. This book was far better received than I could have dreamed. Most humbling were letters I received from climbers who told me that it helped them not just with climbing but with overcoming addiction or other significant issues that affected their quality of life. Like me, they leaned on climbing as a means to cope with other problems in life, but found that the same tools to optimise training were also useful in addressing those other problems as well.

However, *9 out of 10...* left two glaring gaps, one of which I hoped to fill with this book. In *9 out of 10...* and my blog, I wrote from the perspective of an experienced coach and qualified sports scientist. On the blog, I tended to write about current climbs to illustrate my own philosophy and practice of training, since I continued to improve after climbing Rhapsody. I have not written much about my long apprenticeship in climbing, or about the mistakes I learned from during that process. Moreover, it was only with hindsight that I grasped the importance of a few fortunate events during that apprenticeship; decisions made by accident, attributes I inherited from my parents

On the first ascent of Axiom (8a), Tunnel Wall, Glen Coe in 2005. Scotland maintains a fine balance between bolted sport climbing and bold traditional climbing with natural protection. Tunnel Wall exemplifies the border between the two disciplines. I first inspected this line as a very bold trad route, around E9. The more I looked at it, the more it seemed axiomatic that it would be better as a bolted route like the other lines just to the right. The lines on either side of the orange 'rump' of the crag are subtly different in character and make better traditional routes, even though many are very serious with poor protection.
© Cubby Images

or witnessed in others. The size and nature of their influence are more obvious to me now as a much more experienced climber, and as a father of a daughter approaching the age I started climbing.

9 out of 10... was written in 2009, at a cusp in the progression of the sport. Before that, there was a relative poverty of information about how to train or hone the art of performance. Since then, there has been a steady increase in online discourse about training, and today, overabundance is the problem. Some information is superfluous or irrelevant. Some has a kernel of truth but is awkwardly shoehorned into the wrong context, perhaps just for the clicks. Some is overrepresented because it can be easily demonstrated in video or packaged into a saleable product. Some is just false and best ignored. *9 out 10...* is a pretty simple book, intended to anchor what I felt were the key pillars of training and the behavioural tendencies that derail efforts to follow them. In 2009, I thought the book would be rendered irrelevant in ten years; these basic principles would be obvious to everyone. Instead, I feel that its content is needed even more. The overabundance of training information has served to crowd out or distract from the key pillars. One thing I have learned from YouTube is the power of a narrative for drawing attention to important ideas. For this reason, I thought it would be valuable to explore how a few key factors pushed me from a timid, weak and fairly unhealthy 15-year-old to breaking ground in trad climbing via 12 years of highly effective apprenticeship. *Moving the Needle* is essentially an applied version of *9 out of 10 climbers make the same mistakes*.

In recent years, I have become self-conscious about two opposing views that others have sometimes expressed about me. Those who only know me through my YouTube channel sometimes comment in a manner that seems to overestimate my natural ability, especially for bold climbing. I have the capacity to act with boldness, even fearlessness. But I am not inherently bold by nature. I post videos of myself doing one-arm pull-ups on 20mm edges, but do not have videos to show off how weak I was in 1997. I hope that describing the process I went through to reach higher grades will emphasise that the potential level of every climber is very high and that disadvantages in one area can lay the foundations for progress in another, if circumstances are right.

On the flip side, folk who know a bit more about me will be aware of just how poor my attributes in certain areas of climbing can be. For example, I still perform very poorly in warm conditions and to this day struggle to maintain an average level on indoor walls. Summer is still my least favourite season, and I retreat as high as I can to find cold conditions. Unfortunately, I have not found a satisfactory way to reduce the sweating in my fingers other than seeking wind and cold. I've drawn attention to this several times in this book and I hope it underlines how significant tactics are in climbing and how useful it can be to figure out how to make the most of strengths you do have and turn weaknesses to your advantage.

I hope that observers from either perspective will be heartened to see just how low my standard was, how long it took me to learn key lessons and that taking a long time to do so may have helped me reach further than might have been expected. In other words, it is a lament to an old-school climbing apprenticeship. Dumby Dave was a nickname sometimes used to poke fun at my apparently narrow focus, at least along geographical lines. Stevie Haston once called me 'Britain's best parochial climber'. I wish I could say that I knew in advance how valuable it would later become to delay the widening of my geographical horizons, and focus for longer on deepening them. But it seems important for me to share the benefit of hindsight now.

The second gap left open by my previous writing relates to how to sustain a high standard of athletic performance over time. The length and content of my apprenticeship did not just raise the peak of the level I was able to achieve, it was also key in shaping my onward trajectory. In fact, calling it the peak I was able to achieve may yet be premature. I am 46 years old, and last year I climbed XII mixed, Font 8B+ (V14) boulder and E11 trad. It's the first time I've been able to manage this level across all those disciplines in the same year. Am I still getting better at climbing? It is hard to say for certain. But if I frame the question a different way, I can offer a more confident answer. Is there anything I have previously climbed in any climbing discipline that I could not still climb today? No. I am fairly certain of that.

The subject of maintaining and building on athletic success as age advances is, to me, even more interesting than reaching the heights of youthful performance. I suppose that is to be expected, given my age. But I don't think it's just a personal mission. Nudging the limits of performance as an experienced athlete and doing so in the face of life's battle scars is a formidable challenge. I left my undergraduate study of sports science still unclear as to what explains the variation in longevity among athletes across different sports or even within them. I noted that many climbers peaked late compared to athletes in mainstream sports and sustained it with grey hair, replaced hips, and hands that would barely close. How is this possible? Some aspects of the Western lifestyle I observed in poverty-ridden Dumbarton are strangely also common to professional sportspeople: the relentlessness of stress, disturbed sleep, substance abuse and, even for sponsored athletes, exploitation by employers who ask too much and give too little. This sympathetic fight-or-flight-driven lifestyle is incompatible with our physiological makeup if sustained for long periods, and it catches up with you quite quickly.

Right: On the First ascent of Ring of Steall (8c+), Glen Nevis, 2007. This route was a project of Dave Cuthbertson's in 1992. He was very close to completing what would have been one of the first 8c+ routes in the world at the time. Completing it was one of my first targets after Rhapsody and was when I felt a sense of having reached a level of strength and fitness approaching Cubby's.
© Claire MacLeod

Professional sport tends to take the cream of healthy youth, chew them up and spit them out at age 30 in a poor physical and mental state, sometimes glad to see the back of their sport for good. I am sad to say that I now have an even more jaded view of mainstream professional sports than I did at school, and I would strongly caution my daughter against entering that system. When navigated safely, it can be deeply rewarding, but it is easily as dangerous as anything I have described in this book. You need your wits about you to get through it.

Today, climbing is edging towards the mainstream. Large shiny climbing gyms in every city, Olympic champions and well-curated social media channels. With it, there are more frequent and visible examples of young athletes burning out, suffering the consequences of eating disorders, or simply posting on social media in a manner that implies unhappiness simmering below the surface. There are plenty of examples of the opposite, of course. But I hope climbers will always look out for those who are struggling or in danger as much as they admire those on the summit. This combination of training stress and a Western lifestyle remains a relative blind spot in sports

Right: On the first ascent of The Gathering (E7 6b) on the Cioch, Isle of Skye in 2004. Strength and technical ability gained at Dumbarton Rock was a great foundation to take on the countless unclimbed lines around the Highlands and Islands of Scotland. © Cubby Images

science, although knowledgeable and experienced coaches are now pointing out that sympathetic activation via training stress, life stress, nutritional stress and other environmental factors must be balanced with parasympathetic activation to absorb, adapt and become more resilient.

Despite the popularity of activities like yoga and meditation in wider culture, I think adoption of these recovery techniques is underutilised by higher-level athletes. Of course, activities that balance the intensity of training and Western life can come in multiple forms. I don't want to end this section without making the obvious point that climbers should always make use of a key source of resilience against the rigours of sport and life: time on mountains. In the early 2000s, I followed the career of Italian all-rounder Mauro 'Bubu' Bole, since he was excelling in the same spread of disciplines I aspired to. He was one of the first climbers to have a website, and its header bar said, 'Every once in a while, I stop, just to look.' At the time, I thought this was a curious quote to choose for the landing page and the first impression he wanted visitors to form. Despite spending lots of time in the Scottish mountains, I was slow to appreciate the

Top right: Looking down the Clyde from the Dumbarton boulders.

Bottom right: With Claire after repeating an E9.

value of time there with no objective other than to be. After climbing Rhapsody and adding more layers of complication to my life, I had to learn the hard way that simply sitting watching trees in the wind, snow fall or the light change on rocks and hills was not separate from my performance as a rock climber but, in part, the basis of it. When I removed myself from it for too long or was too busy to stop on mountains as Bubu suggested, I ran into problems as a rock climber and as a person.

I could have looked back and recognised much sooner that quietly sitting under my little endurance circuit at Dumbarton Rock just watching the river was as important as the circuits themselves for the jump in standard that allowed me to climb E11. I failed to appreciate that resting here was not the same as resting between circuits in a climbing gym. Even now I do appreciate the difference, I still veer towards a sympathetic-dominant lifestyle and I owe much to Claire for constantly reminding me to properly rest and replenish my capacity to push hard again. I also admire her for having the courage to be unapologetic about leaving time and space to look after her own recovery in a culture that does not understand the value of it. I would like to explore this fascinating and important subject of sustaining high performance in climbing in future books.

Earlier in this book, I introduced my struggles with moderate depression and a tendency for weight gain, something I share with a large proportion of the population of my own and most other Western countries. But I have left that story hanging because in the period this book covers, those issues remained unresolved. I managed the symptoms of both fairly successfully, even if I could not cure them. To be honest, I did not imagine anything more than symptom management was possible. I do think it is important to note that I was able to achieve a world-class level in my sport while suffering from both conditions (along with others I haven't mentioned, such as severe eczema). Consistent movement, time on mountains, connection with people who treated me as equal and a stable home environment all helped me *manage* depression. Restriction of junk food and constant attention to calorie intake roughly controlled the symptom of weight dysregulation that people pay most attention to: scale weight. But it did not successfully control the constant hunger and effort required to override my appetite.

By 2014, fighting my body to stop excess weight gain was becoming tiring, and I was driven to study nutrition and answer a suspicion that I was missing something. By that point, I had been a professional climber recognised for a disciplined approach for long enough to doubt the notion that I was unable to exert conscious control over this one area. My observation at age 16 that hunger actually decreased during a multi-day fast

Right: Wrapping my feet with ointment-soaked bandages after a severe eczema outbreak in my flat in Dumbarton. Wearing rock shoes and mountaineering boots was always painful. Using socks with my rock shoes took the edge off the pain and soaked up blood. After an outbreak, I had to take three days off climbing and bandage my feet to try to let open sores heal.
© Claire MacLeod

had been niggling away at me. I discovered that, like the field of training, the weight management field has many remaining uncertainties, with no consensus on the upstream causes of people's failure to maintain 'calorie balance'.

My study led me to experiment with many different diets, with quite dramatic results across all the remaining weaknesses I mentioned: weight, mood and skin health. I discovered that well-formulated diets rich in vegetables tended to exacerbate all three conditions, with large swings in energy levels and dependence on readily available snacks to avoid energy crashes. Conversely, a diet of only animal foods put my lifelong eczema into remission in two days, and symptoms of depression resolved completely in weeks. It was a similar result with excess weight, which I have been able to control effortlessly since. I would like to give the controversies of nutrition a comprehensive airing in a dedicated future book. Here, I would just like to emphasise three simple points about nutrition that I am fairly confident about and would tell my younger self if I could go back in time. Had I known them as a young adult, I think my story may have read quite differently. First, the type of food you eat is likely to dictate the amount you eat over the long term. Related to this, while it works acceptably well for some, the mantra of 'everything in moderation' may be grossly counterproductive for others. I broke myself against this advice for too many years and hypothesise that it should not be generalised. Finally, animal-derived foods appear to be grossly underrated in health and performance nutrition.

Whether you are a rock climber or not, I hope this book has drawn attention to the many layers that climbing can offer. I started by opening an atlas, triggering a fleeting notion to explore. The interest and reward I got from mountains then nurtured and engrained that impulse, which later expanded beyond the physical world of cliffs to include finding the limits of my own performance and helping others do the same. Exploration of new climbs in far-flung places around the world is a fantastic thing to do, and after my apprenticeship, I broadened my targets to mountains in the Alps, Norway and Patagonia. But I hope I have shown that climbers can also explore depth as well as breadth by focusing on the complexity of movement and limits of human physiology on cliffs or mountains which are just down the road. I now find my focus returning to the endless unclimbed lines in Scotland to satisfy my curiosity about how to move the needle and open up new possibilities. I also find myself removing complexity from my life again, realising that Scottish mountains and curiosity are all I need for a deeply adventurous life.

THANKS

I would like to thank Claire and Freida MacLeod for their patience and support while I wrote this book. Deziree Wilson carefully edited this book, and I thank her for her invaluable help in teasing out the important elements of my experience.

For their great photography, I thank Dave 'Cubby' Cuthbertson, Steven Gordon, Paul Diffley, Dave Brown, Peter McGowan, Tim Morozzo, John Watson, Steve Richardson, Niall McNair, Andy Turner, Nick Tarmey, Andy McIntyre, Scott Muir, Richard McGhee and Alasdair Weir. Thanks to Barbara MacLeod for early pictures of climbing and counsel on how to present difficult elements of the story. Thanks also to Katy MacLeod, Calum Muskett, Kevin Woods and Ramon Marin for their valuable input. Tim Parkin very generously helped me with scanning and editing slides.

I'd also like to thank everyone who commented on my blog, YouTube channel etc. over the years, either to encourage me to keep writing about climbing or to ask great questions that offered a new perspective about the activty we share and love.

OTHER BOOKS BY DAVE MACLEOD

9 OUT OF 10 CLIMBERS MAKE THE SAME MISTAKES

The bestselling book on improving your climbing performance. *9 out of 10...* sharpens your focus on the key pillars of climbing performance, eliminates the unimportant and offers behavioural tools to stay on the improvement path that so many climbers struggle to follow.

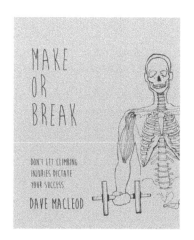

MAKE OR BREAK

Sooner or later, every climber gets injured. Prevention and treatment of finger, elbow, shoulder and other climbing-related injuries end up as the limiting factor for enjoyment and progression. *Make or Break* highlights the key priorities for building resilience against injury and summarises the current research from sports medicine and physiotherapy for building successful rehabilitation protocols.